"I am very impressed by the monograph *Scientometrics for the Humanities and Social Sciences* by R. Sooryamoorthy. It is the first monograph on Scientometrics I am aware of that is devoted exclusively to the use of scientometric techniques in the study of the humanities and social sciences. It aims to unfold the potential of these techniques to a wide audience. A most interesting and unique feature of the book is the well-designed presentation of a large number of cases illustrating both the methodology itself as well as its application in the study of the development of humanities and social sciences and its numerous subject fields, and in the assessment of research activity and performance. Most importantly, it dedicates attention not only to the potential of these techniques, but also to its limitations and pitfalls. This makes the monograph a valuable and useful piece of work, unmissable for librarians, research managers and policy makers, and researchers in the domain of humanities and social sciences who aim to enlarge their understanding of the uses and limits of scientometric techniques in these domains of science and scholarship."

Henk F. Moed, Formerly at the Centre for Science and
Technology Studies (CWTS), Leiden University, the Netherlands

"Scientometric analyses seem straightforward given the availability of apparently clear metrics (on production, citation and other operations of science). And indeed, measurement systems and underlying data collections are continually developing to cope with various difficulties they still face. But, whereas the use of such metrics works quite well for natural science, their application to the social sciences and even worse for the humanities is bedeviled by a further array of problems. Prof. Sooryamoorthy's book takes us through such issues, placing them in historical context. This guide will be invaluable to neophyte analysts wishing to measure social science."

Charles Crothers, Auckland University of Technology, New Zealand

"*Scientometrics for the Humanities and Social Sciences* provides a compendious overview of the origin of scientometrics and is a perfect collection of illustrative case studies that can guide social scientists to explore their own research fields."

Valeria Aman, German Centre for Higher Education Research
and Science Studies (DZHW), Germany

ii

SCIENTOMETRICS FOR THE HUMANITIES AND SOCIAL SCIENCES

Scientometrics for the Humanities and Social Sciences is the first ever book on scientometrics that deals with the historical development of both quantitative and qualitative data analysis in scientometric studies. It focuses on its applicability in new and emerging areas of inquiry.

This important book presents the inherent potential for data mining and analysis of qualitative data in scientometrics. The author provides select cases of scientometric studies in the humanities and social sciences, explaining their research objectives, sources of data and methodologies. It illustrates how data can be gathered not only from prominent online databases and repositories, but also from journals that are not stored in these databases. With the support of specific examples, the book shows how data on demographic variables can be collected to supplement scientometric data.

The book deals with a research methodology which has an increasing applicability not only to the study of science, but also to the study of the disciplines in the humanities and social sciences.

R. Sooryamoorthy is Professor of Sociology at the University of KwaZulu-Natal, South Africa. A scientist accredited to the National Research Foundation, he is the coauthor of *Science in Participatory Development* (1994) and author of *Transforming Science in South Africa* (2015) and *Science, Policy and Development in Africa* (2020).

SCIENTOMETRICS FOR THE HUMANITIES AND SOCIAL SCIENCES

R. Sooryamoorthy

Routledge
Taylor & Francis Group

LONDON AND NEW YORK

First published 2021
by Routledge
2 Park Square, Milton Park, Abingdon, Oxon OX14 4RN

and by Routledge
52 Vanderbilt Avenue, New York, NY 10017

Routledge is an imprint of the Taylor & Francis Group, an informa business

British Library Cataloguing-in-Publication Data
A catalogue record for this book is available from the British Library

Library of Congress Cataloging-in-Publication Data
A catalog record has been requested for this book

ISBN: 9780367626860 (hbk)
ISBN: 9780367627010 (pbk)
ISBN: 9781003110415 (ebk)

Typeset in Bembo
by Newgen Publishing UK

To
My son Dakshin

CONTENTS

FIGURES

CASES

PREFACE

I do not remember when the idea to propose a book like this came to mind. But my intention was to revive the interests of scholars in scientometrics and to attract students and scholars to undertake scientometric studies in the humanities and social sciences (HSS). This is the rationale of a book that will show the importance of scientometrics, the potential in terms of data sources and the procedures to conduct solid scientometric studies in the HSS. The focus of the book, however, should not prevent students and scholars in science from using the book for their fields of interest.

For any newcomer to the field, or an aspirant, the book is of value and help. It describes where and how to start with a scientometric study, drawing from the world of rich and expansive data that has already been collected and stored. It is hoped that the book will contribute to the development of more creative and innovative scientometric studies in the coming years. In this book, the approach is to develop a sustained interest in the HSS and the applications of scientometrics to the study of the HSS.

The production of a book involves the efforts of many. I am grateful to the colleagues who work in this specialised area for their wisdom, my university for granting me a sabbatical, Sandra Waters for suggesting changes and corrections, and Hannah Shakespeare, the senior commissioning editor of research methods at Routledge and her assistant Matt Bickerton. Working with Hannah was a memorable experience. I commend her for her prompt communication and professional efficiency. The support I received from my wife was remarkable. My son Dakshin gave me several wonderful moments in my life that made me proud of him. I dedicate this book to him, a token of my love and affection for him.

ABBREVIATIONS

A&HCI	*Arts and Humanities Citation Index*
ASEAN	Association of Southeast Asian Nations
BkCI	*Book Citation Index*
BkCI-S	*Book Citation Index – Science*
BkCI-SSH	*Book Citation Index – Social Sciences and Humanities*
CMP	Condensed matter physics
DOI	Digital object identifier
ESCI	*Emerging Sources Citation Index*
GBS	Google Books search
GS	Google Scholar
GSC	Google Scholar Citations
GSM	Google Scholar Metrics
HSS	Humanities and social sciences
ISI	Institute for Scientific Information
ISSI	International Society for Scientometrics and Informetrics
JCR	*Journal Citation Reports*
JIF	Journal impact factor
SCI	*Science Citation Index*
SciELO	Scientific Electronic Library Online
SPSS	Statistical Package for the Social Sciences
SSCI	*Social Sciences Citation Index*
STEM	Science, technology, engineering and mathematics
WoS	Web of Science
WoSCC	Web of Science Core Collection

ABOUT THE BOOK

Since its origin in the late 1960s, scientometrics has grown and flourished. It has become a recognised field of study and has a set of methods which can be used in a host of areas of communication in both science and non-science disciplines. By any standard, this was not a mean achievement. Scientometrics is acclaimed to be an effective tool for the study of research assessment, evaluation and performance, collaboration, citation and impact. Studies of this nature were done at levels varying from micro to macro. Individuals, institutions, countries and scholars could be evaluated and assessed using reliable scientometric indicators which have become meaningful instruments in the study of scientific disciplines.

Institutions of higher learning and governments adopt scientometric measures to inform their decisions and to develop policies pertaining to ranking, standing, funding, impact, visibility and future plans. In a way, scientometrics serves the purpose of determining the future of academic disciplines and scientific research. With the application of scientometric tools, the need, demand and necessity of science and non-science can be measured and assessed. The uses and applications of scientometrics continue to evolve, unfolding and embracing new areas of scientific inquiry.

Over the last few decades, scientometrics has attracted the attention of a growing scientific community, which is seemingly excited about its potential, possibilities and capabilities. Academic journals devoted to scientometric research were launched while other academic journals welcomed scientometric studies. Research centres and institutes conducting scientometric research have also been established in different parts of the world. This is a testimony to the value scientometrics was bringing to its trusted applications.

In its early years of origin and development, scientometrics was largely centred on the sphere of science and scientific disciplines. Its uses and applications in the humanities and social sciences (HSS) have not been as common and popular as in science. This restricted scientometrics from expanding its horizon to new avenues of inquiry in the HSS in particular. Scholars have now found scientometrics appropriate for the study of disciplines in science or in the HSS.

Given the diminishing importance of the HSS in relation to science, as evident in the dwindling support and funding received by HSS disciplines, scientometrics would have been a saviour of the HSS. Being a widely used tool for research assessment and mapping, scientometric studies could have helped the HSS disciplines in unpredictable ways. Scientometric studies could have shown the relevance of disciplines and provided directions to refocus where necessary. Knowledge of disciplinary features, emerging or dying areas of knowledge within disciplines, authorship patterns, collaboration or lack of it among authors, methodological and theoretical developments, referencing patterns, citations and impact are crucial for the existence, directions and survival of the disciplines in the HSS. Information gained through scientometric approaches would have influenced those disciplines that needed to chart new territories, fields, methodologies and theories.

The insubstantial volume of scientometric studies in the fields of the HSS did not augur well for either the disciplines or for their scholars. This is a missed opportunity for both the disciplines and the scholars who are part of the HSS. Why is this happening? When social scientists are mostly the ones who conduct scientometric studies of science, why are they indifferent to their own parental disciplines? They have a commitment to the growth and development of their disciplines. Should the HSS continue to be neglected by its own family as we see today? Do they deserve better consideration from their own members of the disciplines? Why do social scientists shy away from studying their own or sister disciplines? These questions led to the origin of this book on scientometrics.

While textual analysis is not alien to scientometric studies, the mining and use of qualitative data has not been significant. This is a major drawback of scientometrics. Large quantities of qualitative data stored in the citation indexes, journal websites and online bookstores remain untapped for fruitful scientometric analyses. For the HSS, this is an indispensable and invaluable source of information. The presence of qualitative data will also dispel the notions among the scientific community that scientometrics is just a set of quantitative methods, techniques, measures and indicators. This notion has also prevented students and scholars from entering the field for want of quantitative methodological skills.

The focus of the book is largely on scientometrics in and for the HSS, which distinguishes itself from other books on scientometrics. But it is inevitable to show which scientometric studies have already been undertaken and published. While not following a standard review-of-literature format, relevant studies have been carefully chosen as illustrative cases and examples. A judicious selection of cases and examples is made to represent the core areas of scientometrics such as mapping, research evaluation, research performance, collaboration, citations and impact. Sixty-four cases and examples are examined for their research objectives and questions, methodological procedures, sources of data, analysis strategies and findings.

The cases and examples described in the book are also intended to serve as reference models for the community of students and scholars to critically examine and adopt in their studies when necessary. This is one way to learn and practise a methodology. They also illustrate how creatively and innovatively scientometric data can be employed, enriched and developed. Current practitioners of scientometric methods will find these cases and examples adaptable to whatever modifications they may require.

It is the purpose of this book to illustrate how the major databases can be successfully utilised for studying the fields, subjects and disciplines of the HSS. The explanations of the applications of these databases in the book serve as models and examples that can be repeated in other lesser-known databases and are available in languages other than English. With the addition of supplementary data such as that pertaining to gender, race, age and rank of authors, scientometric data can be improved for more powerful analyses. This is rarely done in scientometrics. The book explains how to develop a more useful dataset capable of producing robust findings.

The treatment of scientometrics concerning its applications is two-fold in this book. One, the book presents the methods and applications of scientometrics to the fields of science which encompass the sociological, economic, historical and developmental aspects of science. Two, it shows how the usefulness of scientometrics can be extended to the study of fields and disciplines in the HSS, which is largely underutilised and underexplored. This is where scientometrics as a specific field of study of scientific research can be achieved by expanding its coverage to all scientific disciplines in both science and the HSS.

Chapter 1 deals with the major features of scientometrics, laying the basis for the rest of the book. It covers its origin, development, taxonomy, theoretical foundations and paradigms. This is essential to the understanding of the relevance of scientometrics in the HSS. The purpose of Chapter 2 is to provide a comprehensive view of the current applications of scientometrics. The chapter covers the uses of scientometrics in research evaluation and impact, coauthorship, collaboration, mapping of disciplines and subjects, citation studies and big data. Chapter 3 discusses the potential and challenges for scientometrics. Challenges relating to the typical publication practices in the HSS and coverage of the prospects for strong scientometric studies are highlighted. Chapter 4 presents concrete case studies by relying on specific scientometric studies representing a range of disciplines and subject fields in the HSS. The objective of these selected cases and examples is to identify their methodologies, sources of data and procedures of analysis. The chapter shows how creatively and innovatively scientometric studies can be performed in the HSS.

Some of the challenges of scientometric research relate to the selection of appropriate databases which provide the data sources for scientometric studies in the

HSS. It is often perplexing to choose one as there are issues of coverage and availability. Chapter 5 is devoted to the sources, processing and analysis of data. It critically examines and compares data sources, their advantages, reliability, challenges and limitations. One of the drawbacks of scientometric studies is that they have not utilised the rich resources of qualitative data that are available in citation indexes and publication platforms. How such valuable qualitative data can be obtained for scientometric studies is shown in this final chapter.

1

SCIENTOMETRICS

An introduction

Introduction

Measuring scientific research and its impact is a serious exercise that the stakeholders in the domains of higher education, science, technology and innovation undertake regularly. When the funding, ranking, image and reputation of scholars and institutions are concerned, research assessment is an inevitable requirement. In this assessment, which is often initiated by individuals, institutions, organisations or governments, several tools are adopted. Among them are research outputs such as publications and patents, scientific methods and indicators. Researchers, universities, research institutions, organisations and countries alike adopt these tools for assessing research. It is expected that such tools should work well for making informed decisions that eventually lead to appropriate policies. Therefore, it is pertinent that proper and effective evaluative measures that can gauge the impact of scientific research with accuracy and quality are applied.

Scientific measurements and indicators have gained value in the overall assessment and evaluation of research. They are also crucial for an institution or a country to chart its own course of scientific advancement. Appropriate policies are to be based on the information gathered through reliable scientific measures and methodology. With the help of measurable quantitative indicators, it is now possible to evaluate and assess professors, researchers, training programmes and universities (Gingras, 2016).

Scientometrics has matured into a field of research that has seen refinements, reworkings and enhancements, but without any fundamental changes to its taxonomy (Bornmann, 2014). The refinements are evident on several fronts. The indicators used in scientometrics for research evaluation are normalised for subject category and year of publication; co-citation analysis is used to identify research fronts; and the analysis of coauthorships is meaningful to find collaboration between research groups, universities and countries (Bornmann, 2014).

Countries have developed suitable models for research assessment and evaluation. The Norwegian model, to cite an example, is being used by countries such as Belgium, Denmark, Finland, Norway, Sweden and Portugal for decision making in research funding and research allocations (Sivertsen, 2016). The model takes into account the structured, verifiable and validated peer-reviewed scholarly literature in all research areas, including the humanities (Sivertsen, 2016). The significance of a reliable methodology like scientometrics to assess scientific research in both science and the humanities and social sciences (HSS) comes into focus at this juncture.

In order to understand scientometrics and its special significance in the HSS, knowledge about some relevant aspects of scientometrics is necessary. The sections that follow discuss the origin, development, taxonomy, theoretical foundations and scientometric paradigms. Being the central focus of the book, the chapter concludes with a discussion on scientometrics for the HSS and thus sets the background for scientometrics and its applications in the HSS.

Origin and development

Over the years, scientometrics has evolved into an effective tool for the assessment, valuation and impact of research. It has demonstrated its ability to offer reliable, transparent and relevant results (Bornmann and Leydesdorff, 2014). Scientometrics has proved to be an extremely useful tool in parsing the vast and complex system of science, and in analysing vast amounts of data (Sugimoto and Larivière, 2018). Viewed either as a field of study or a collection of methods, scientometrics is not meant for the above research objective alone. It is but only one of its several applications.

The word metrics in scientometrics connotes the fields from which it emerged and its use of mathematical methods (Zhang et al., 2013). Consisting of multiple methods, scientometrics was initially and mostly applied to the study of science and scientific knowledge. Scientific research produced in the realm of the HSS was not part of the original concern of scientometrics. It was deemed to be just a metrics of science, studied mostly by social scientists. Gradually and steadily, it grew into a strong scientific field with a host of methods that found wider applications beyond the confines of science. This becomes clearer as the following more detailed discussion shows. Even professional historians of science paid very little attention to the social sciences (Lazarsfeld, 1961).

Scientometrics is about the science of communication, which is performed through scientific publications, people and institutions. In order to measure science, one needs to measure the process of scientific communication which, in very simpler terms, is a network of publications (nodes) connected by citations (directed links) (Szántó-Várnagy et al., 2014). Scientometrics in this sense is a set of methods to study scientific and academic publications. It does, however, have a clear focus on the producers, processes and evolution of research, using publications as a proxy to research (Belter, 2015).

Early developments

Since the early 20th century, scholars have been taking the study of scientific publications seriously. Attempts were made to examine publications in a more systematic and scientific way. In one of the earliest studies of publications in 1906, James McKeen Cattell, a psychologist, compiled a biographical directory, *American Men of Science*. He had been producing statistics about scientists and their geographical distribution regularly (Godin, 2006, 2007). Cattell's efforts were soon to be conceded as the first step towards systematic efforts for the quantitative studies of science (Godin, 2007), which he accomplished through the analysis of publications. To his advantage, Cattell was the editor of the journal *Science* during 1895–1944, which gave him the chance to study publications. While he was preparing the directory *American Men of Science*, he did not realise that he was employing a method which later came to be known as scientometrics.

Shortly after this, in 1927, Paul L. K. Gross and E. M. Gross conducted a study of publications in chemistry. This was to find out the crucial periodicals that were indispensable to the academic purpose of students (Gross and Gross, 1927). They tabulated the references in the *Journal of the American Chemical Society* to assist students who were doing chemistry. Viewed as a pioneering step in citation analysis, the far-reaching consequences of this by the husband and wife were to be seen in citation analysis in the years to come (Andrés, 2009; Glänzel, 2003). Gross and Gross were the first to consider the structure of academic literature and citation count to make an evaluation (Hertzel, 2003).

In this inception stage, endeavours were intended primarily to manage journal publications and references cited in publications to assist students, professors and researchers. Based on the count of citations, librarians were able to separate the journals that were obsolete, as they were rarely cited, and those that could be useful for students and researchers (Gingras, 2016). At this point, the study of publications was influenced by the same or related contributions from several others.

To study the history of science

A British physicist, John Desmond Bernal (1901–1971), published his book *The Social Function of Science* in 1939. Bernal held that science will be recognised as a chief factor in fundamental social change, and the relation and interaction between science and society cannot be neglected. He emphasised that science is an integral part of the material and economic life of our times and argued for the application of scientific methods to its study (Bernal, 1939). *The Social Function of Science* attracted a great deal of interest from scholars and public alike. People began to accept that the production of science is as important as its impact on society. The knowledge about the history of science was therefore indispensable. It became necessary to study the production of science and how science is communicated to the outside world.

George Alfred Léon Sarton (1884–1956), a Belgium-born American chemist, mathematician and historian, published his voluminous three-volume *Introduction to the History of Science* over a period of 21 years during 1927–1947 (Sarton, 1927, 1931, 1947). It was then a major work on the history of science. Sarton's other related contributions, such as *A History of Science* (1952 , 1959) and *The Study of the History of Science* (1936), were equally exceptional. His commitment to the study of science led him to start two academic journals devoted to the philosophy and history of science. All his efforts made him recognised as a pioneer in establishing the history of science as a discipline (Garfield, 1985). For him, as he repeatedly emphasised in his multiple works, the history of science is the only history that illustrates the progress of mankind.

It was around this time, in 1935, that two Polish scholars, María Ossowska and Stanislaw Ossowski, philosophers and sociologists, drew the attention to the study of scientific research. Later appearing in English (Ossowska and Ossowski, 1964),[1] their work explained why science, including its research activities and the products of these activities, should be a subject of scientific investigation. To them, the science of science falls into three groups of study of problems related to science and scientists. They are connected with the personality of the workers in science; the activities leading to the formation of science; and science as a completed human product.

The contributions of these stalwarts, and several others who were either their contemporaries or followers, gave rise to an apparent increase in a genuine interest in science and its social function. In order to understand science and its interaction with society, it was necessary to know what science is, how it is produced, what it is contributing and what effects and impact it has on both science and society. The scientific study of science was gradually gaining momentum.

Following the inputs of these pioneers and the effects they made on the realm of science and the study of science, more significant publications appeared. Not long afterwards, the key works of two prominent figures, Eugene Garfield (1925–2017) and Derek John de Solla Price (1922–1983), were published. In 1955, Garfield, an American linguist, produced his work on citation indexes in the journal *Science*. This was later to become the cornerstone of the study of science using citations. The publication paved the way not only for the study of science, but also for the study of subjects in the HSS. Through this publication, Garfield introduced the idea of documentation. Within a few years, Price, a British physicist and mathematician, published his classical work, *Little Science, Big Science* (1963). These two major contributions signalled the genesis of scientometrics.

Emergence of citation indexes

Using published scientific literature, Garfield pioneered citation indexes for the study of science. A citation index is a database that contains a large collection of records of publications, called bibliographic records. These records contain information about publications such as journal articles, reviews, book chapters, monographs, edited volumes, conference proceedings, notes and reports.[2] The metadata of the publication records consists of the type of documents, names and institutional address of authors, the year of publication, title of publication, title of journal/book, publisher's details, page length of the publications, citations and references cited in the publication.

To Garfield, citation indexes were crucial to the study of scientific knowledge. He began with the study of science, but did not neglect the HSS and their relevance. Robert Merton, who is known for his works on the sociology of science, once asked why the *Science Citation Index* (*SCI*) was not introduced until 1963, suggesting that the use of the *SCI* by sociologists of science was only a by-product (Broadus, 1987). This was due to the attention of the methods in scientometrics that were believed to be applicable only to the sociology of science. As prolific an author as he was, Garfield dwelt on many other aspects of science. His numerous publications in the field are notable for a scholar to achieve in a lifetime. Garfield's influence continues in scientometrics.

The influence of Garfield's efforts on the need for an advanced citation index meant for the study of science was huge in the subsequent years. His intention to develop a citation index, or an association-of-idea index as he preferred to call it, was fundamentally to eliminate the flaws in scientific literature. By making researchers aware of the criticisms of previous papers, Garfield was seeking to flag the uncritical citation of fraudulent, incomplete or obsolete data (Garfield, 1955). In this association-of-idea index, Garfield conceived the units of thought. A book can act as a macro unit of thought while a journal article is a micro unit of thought; and an index deals in the submicro or molecular unit of thought. Garfield was convinced that a citation index would be handy for researchers to check the previous publications in their area of interest (Garfield, 1955). This again makes the case for scientometrics in the study of scientific knowledge in all fields.

With contributions to the study of science, Derek John de Solla Price was to become a historian of science. His knowledge of mathematics and statistics equipped him to make a substantial improvement towards quantification in the social studies of science (Garfield, 1992). Price's work, *Little Science, Big Science,* offered a systematic approach to the study of the structure of modern science, and laid the foundations for the study of research evaluation (Andrés, 2009). Research evaluation was soon to be a core area of scientometrics.

As early as 1946, the Royal Society of London organised two international conferences on scientific information: The Royal Society Empire Scientific Conference, and the British Commonwealth Official Scientific Conference. These conferences underlined the need for adequate scientific information services (Ball, 1950) for the benefit of science and researchers. Both conferences brought together libraries, societies and institutions engaged in publishing, abstracting and information services. At these conferences, the possibility of an improvement of the existing methods of collection, indexing and distribution of scientific literature was deliberated (Ball, 1950).

Two years later, in 1948, a more expanded conference named The Royal Society Scientific Information Conference, was to take place in London. The conference witnessed the agglomeration of prominent figures such as J. D. Bernal, J. E. Holmstrom and E. N. da C. Andrade, among others. Issues pertaining to scientific publishing and techniques and quality of abstracting were examined at the conference.

Conceptualising scientometrics

The initial steps were later conceptualised and developed into scientometrics. However, it took a few more years before the term scientometrics was actually born. In 1969, the Russian mathematician-philosopher Vassily Vassilievich Nalimov (1910–1997) coined and defined the term scientometrics. Nalimov defined scientometrics (*naukometriya* in Russian) as the application of quantitative methods for the analysis of all kinds of scientific activities in all disciplines (Nalimov and Mulchenko, 1969). This definition nevertheless did not distinguish between science and non-science subjects such as the HSS.

Little Science, Big Science … and Beyond[3] contained one of Price's famous papers, "Networks of scientific papers", which, according to Garfield and Merton, is the most important single contribution to information science. In this paper, Price (1986) showed that "the patterns of citation to papers composing that literature define the parameters of research fronts in science". This triggered scientometric research on emerging research areas and fields in disciplines. The network of citations in later years resulted in extensive scientometric studies on networks and collaboration (also influenced by his work on collaboration in invisible college). Networks and collaboration have become one of the most flourishing and fascinating areas of research in scientometrics, not only in science but also in the HSS. This opened up investigations into several other dimensions of scientific research, covering the typologies of collaboration and their relationship to citations, impact, visibility, productivity and disciplinary developments. Collaborative authorship is not unique to science but also applies to the HSS.

Price's thinking was along the lines of using the tools of science to the study of science. He asked why we should not use the tools of science to science, and measure, generalise and formulate hypotheses to draw conclusions. Price argued for a quantitative approach since there was a large corpus of numerical data on manpower, money, publications and institutions available, as well as administrative situations that called for quantitative decisions (Price, 1965).

Price (1963, 1986) contended that science grows at compound interest, multiplying by some fixed amount in equal periods of time. The rate of growth of science at any time is proportional to the size of the population or to the total magnitude already achieved, and the bigger something is, the faster it grows. He desired to have a respectable academic discipline that would do for science as economics does for the economic life of nations. It is for sociologists to be knowledgeable about things that are important to society and not for the natural scientists to turn their own tools of trade upon themselves. Price was seeking out a theory that helps to understand the machinery that makes science act the way it does, and grows the way it grows over a period of time.

The 1963 book by Derek John de Solla Price, *Little Science, Big Science*, his several other books like *Science Since Babylon* (1961[1975]) and over 200 papers earned him the honour of the father of scientometrics (Garfield, 2007). In a joint foreword to *Little Science, Big Science*, Eugene Garfield and Robert Merton maintained that the book had crystallised a new element in the historiography and sociology of science, and that it laid the foundations of the field that has come to be recognised as scientometrics. In this foreword, they crowned Derek John de Solla Price as the father of scientometrics.

Unsurprisingly, many others were attracted to scientometrics, drawing inspiration from Price and his influential works. Their contributions advanced the understanding of science and the development of scientometrics.

Toward a Metric of Science (Elkana et al., 1974), the published proceedings of a conference held in Stanford in 1974, served as the most important source of inspiration for scholars to venture into scientometric studies (De Bellis, 2009). The conference saw experts presenting the historical, sociological and economic aspects of the measurement of science, and working towards a coherent model for the development of science indicators (Gingras, 2016).

Scientometric journals and centres

The launch of three journals around the same time, *Research Policy*, *Social Studies of Science*, both in 1971, and *Science and Public Policy*, in 1973, was another landmark in the development of scientometrics. Devoted to the analysis of the social and economic factors in the development of science and technology, these journals provided the required thrust to the study of science (Gingras, 2016). The journals carried scientometric studies but were not limited to science alone.

In 1978, the first Western centre of excellence, the Information Science and Scientometrics Research Unit, was established at the Library of the Hungarian Academy of Sciences in Budapest. In the same year, Tibor Braun, the director of the Centre, introduced the journal *Scientometrics* devoted to scientometrics. Since then, the journal has given a boost to scientometric studies. The Centre functioned towards shaping the field of scientometrics into a new-born scientometric paradigm as we know today (De Bellis, 2009).

Consequently, a number of journals have emerged and accepted scientometric studies for publication. These include: *Collnet Journal of Scientometrics and Information Management*; *Research Policy*; *Journal of Informetrics*; *Journal of the American Society for Information Science and Technology*; *Journal of Documentation*; *Journal of Informetrics*; *Journal of Information Science*; *Cybermetrics: International Journal of Scientometrics*; *Journal of the Medical Library Association*; *Information Processing & Management*; *PLoS One*; *Research Evaluation*; *Social Studies of Science*; and *Information Visualization*. Discipline-specific journals also encouraged authors to submit scientometric papers that deal with the disciplines, journals and subject areas, providing introspective perspectives.

In the 1970s, the study of science was developing into a speciality. Studies of science began to appear in volume. Heightened interest in studying research from a management perspective remained unabated. In the 1980s, two more institutes emerged which played an instrumental role in the development of scientometrics: the Science Policy Research Unit at the University of Sussex in the UK, and the Research Policy and Science Studies (later the Centre for Science and Technology Studies) at the University of Leiden in the Netherlands. The Leiden institute grew into one of the prestigious institutes known for scientometric studies and the development of scientometric indicators. It was during this period, i.e., in the 1980s, that scientometrics evolved into a distinct scientific discipline, with a specific research profile and subfields (Glänzel, 2003). The availability of the databases in machine-readable form, and the development of computer science and technology, aided the development of scientometrics (Glänzel, 2003).

As an accumulated consequence of all these developments, opportunities were increasingly opened up for scholars who were engaged in scientometrics to meet each other and present their research papers at conferences. The International Conference on Bibliometrics and Theoretical Aspects of Information Retrieval, the first conference of this kind, was held in 1987 in Diepenbeek, Belgium. Organised by Leo Egghe and Ronald Rousseau, both scientometric researchers, the conference expressed the interest of scholars in this area. Following this, a series of conferences were held every two years. At the Berlin conference in 1993, a professional association named the International Society for Scientometrics and Informetrics (ISSI), was born. The conference, with a change in name to The International Conference on Scientometrics and Informetrics, became a regular biannual event of the association. Scientometrics had moved from its position as the "Little Scientometrics" to "Big Scientometrics", to borrow the term from Glänzel and Schoepflin (1994).

Sociology of science and the study of social sciences

Scientometrics or its development is incomplete without a discussion on the sociology of science. The sociology of science has branched out into scientometrics and science and technology studies (Sugimoto and Larivière, 2018). In the distinct branch of the sociology of science, the applications of scientometrics are widespread. Robert Merton, the founding father of the sociology of science, himself applied and encouraged the use of information that can be sourced from citation indexes for the study of science. Merton was one of the enthusiastic users of citation indexes such as the *Social Sciences Citation Index* (*SSCI*), produced by the Institute for Scientific Information (ISI). In an interview, Merton spoke about the intrinsic uses of the citation indexes for the social sciences disciplines and their disciplinary contents and boundaries thus:

> The intensive use of the *SSCI* creates a kind of new contact with continuities and discontinuities in recent thought in the social sciences, and also provides

a kind of imagery of the interrelationships between ideas or techniques that ordinarily appear in isolated spheres of work within the larger field of sociology. It gives most students a concrete and specific sense of how ideas spill over into neighbouring disciplines or quasi-disciplines. The *SSCI* alerts them to the ways in which sociological ideas do overflow their disciplinary banks. Students using the *SSCI* can't avoid acquiring a sense of the actual operating texture of disciplines, the reciprocal uses of common materials, ideas, and instruments. This is by all odds the most important aspect of the general training that the *SSCI* provides.

(Garfield, 1975: 244)

In this interview, Merton was suggesting the scope of such information in the *SSCI* that can be extended across social sciences disciplines. The launch of a citation index for the social sciences prompted scholars to look at their own disciplines retrospectively, and the prospective future of those disciplines and subject fields.

While scientometrics is about science and metrics, the word science is not deemed to be exclusively meant as the natural sciences, but incorporates all areas of knowledge production (Sugimoto and Larivière, 2018). According to Glänzel (2003), the method currently has a target of areas that determine topics and subareas. They are scientometrics for scientometrics, in which methodology is the domain. Then there is scientometrics for scientific disciplines with a larger and diverse interest group, and is the extension of science information. The last one is scientometrics for science policy and management. This is the domain of research evaluation of which the national, regional and institutional structures of science form part. Glänzel (2003) summarises that scientometrics is linked to research in library science, information retrieval and the sociology of science, and it provides services for libraries, scientific information and science policy. These are clearly not in the realm of science but social sciences. The dimensions and corresponding units of analysis in scientometrics, as Leydesdorff (2001a, 2001b) shows, have various levels of aggregations. As Leydesdorff (2001b: 4) succinctly states below, he is providing an overarching view of the domains of scientometrics, its foci, units of analysis and the potential for the study of disciplines:

> … words are organised in texts, scientific articles in journals, journals belong to archives; scientists compose research groups, research groups belong to scientific communities; knowledge claims are based on theories, theories are embedded in disciplines.

Eventually, scholars became interested in examining their own disciplines, using scientometric data, indicators and methods, which will be seen in Chapters 2 and 4. Scientometrics ceased to be an exclusive field of the study of science, but transformed into a field assimilating all scientific fields of knowledge regardless of disciplinary contents and boundaries. Even the prestigious journal *Scientometrics* has carried papers on social sciences ever since its first volume was published. The papers of Small and Crane (1979, in volume 1), Riecken (1980, in volume 2) and Rushton and Meltzer (1981, in volume 3) are examples. At the first international conference on bibliometrics and theoretical aspects of information retrieval, there were papers (Peritz, 1988; and Yitzhaki, 1988, for instance) that touched on social science subjects.

Scientometrics or other forms of similar metrics has remained within a small community of scientists, librarians, sociologists and historians, and became an academic research field in the 1970s (Gingras, 2016). A wider academic community that includes scholars from science and the HSS now undertakes serious scientometric studies. Among them are also policy makers, planning officials, research managers, administrators, higher education officials, funders and organisations who conduct or benefit from scientometric analyses.

Databases for scientometric studies

Reliable databases of the Web of Science (WoS) and Scopus have come into being for facilitating scientometric studies. Invariably, the findings of the analyses based on these databases are recognised and accepted by the scientific community. It does not, however, mean that these are perfect sources of scientometric data. Neither of them indexes all the journals published in every discipline in the natural sciences or in the HSS. Given the proliferation of academic publications and the growing amount of journal publications, it will be a distant possibility and an unrealistic expectation that all the published research will be covered in any single database. Databases that cover papers carried in local journals also exist. As long as this issue of coverage of all research in all disciplines is not resolved, scientometric studies have no option but to depend on these online resources and individual journal resources. This coverage was one of the issues that did not prompt scientometric studies on the HSS, unlike science.

The adoption of technology in the form of software programs also assists in the development of scientometrics. Several data management programs are available for scientometric analysis. Mobile applications such as the one developed by the University of Granada in Spain, called UGRinvestiga,[4] is a new feat for scientometrics and scientometricians (Torres-Salinas and Robinson-García, 2016).

While scientometrics is advancing with numerous studies and developing new indicators that can measure science and research, concerns have been raised about its future. In 1994, Glänzel and Schoepflin were the first to note that there is a lack of consensus about the basic questions and internal communication, and that there is the question of the quality of scientometric research. As they observe, its great possibilities have not been used efficiently, and it is guided by the immediate interests of science policy. In their view, there should be integrated and interdisciplinary approaches that reinforce the fundamental, methodological and experimental research programmes in scientometrics. In contemporary scientometrics, little has occurred in the theoretical and methodological development.

Taxonomy: Scientometrics, bibliometrics and informetrics
Similarities, differences and focus

In the vocabulary of knowledge, a few terms are often used synonymously or inter-changeably, implying similar meanings. Scientometrics, bibliometrics, informetrics, webometrics, cybermetrics and netometrics, for instance, are found to have received the same treatment. These subdisciplines of bibliometrics, informetrics or scientometrics do not seem to be interlinked by their common interests (Glänzel and Schoepflin, 1994). But, as De Bellis (2009) views, the objective of each of these research areas is to analyse, quantify and measure communication for explanatory, evaluative and administrative requirements. The differences among these, therefore, are in the order of the factors and boundaries of what is being measured (De Bellis, 2009).

Scientometrics is the most synonymously used term with bibliometrics and informetrics. It was originally conceived as a method of studying science communication while bibliometrics dealt with more general information about publications (Glänzel, 2003). Bibliometrics has considerable overlap with scientometrics (Thelwall, 2008). Both are not actually one and the same. Bibliometrics is a subset of scientometrics, and it is limited to the analysis of publications and their properties (Gingras, 2016). But all are about metrics, measuring aspects of scientific research and communication. Scientometrics is about science, measurement of science, mapping of disciplines and research outputs. It is concerned with scientific information, and encompasses all quantitative aspects related to the production and dissemination of scientific and technological knowledge (De Bellis, 2009).

Bibliometrics and informetrics focus more on the general information of publications. Generally, bibliometrics undertakes the counting of books, articles and other publications, citations and their statistical manifestations (De Bellis, 2009) and general information processes (Andrés, 2009). Scientometrics addresses the comparative evaluation and contribution of scientists, groups of scientists, institutions and countries (De Bellis, 2009). While bibliometrics is concerned with the quantitative aspects of publications, scientometrics represents a broader view on scientific information (Vinkler, 2010). Scientometrics was initially confined to the measurement of science and communication and was widely used in Europe while bibliometrics was adopted in North America (Andrés, 2009; Brookes, 1988). Scientometrics obtains data from bibliometric databases. It applies methods to bibliometric databases to measure scientific indicators (Cantú-Ortiz, 2018). Here is where scientometrics is aligned to bibliometrics. Scientometric methods help us to understand bibliometric databases in many different ways. Included are the study of scientific citations; scientific impact of researchers; journals and institutions; scientific disciplines; research assessment; research polity and research management; research collaboration; research trends and funding (Cantú-Ortiz, 2018).

The overlaps and distinctiveness of scientometrics and bibliometrics are clarified in a comparative study of scientometrics, bibliometrics and informetrics by

Siluo and Qingli (2017), which throws light on their distinctive aspects. Siluo and Qingli adopted a keyword search of the three metrics in core journals. The usage and distribution of these terms on the basis of the number of publications and cooperation, the research contents, structure and field topics, were examined. The search revealed the position, scope and contents of scientometric, bibliometric and informetric studies. During the period of analysis (2007–2016), the annual volume of publications on bibliometrics was higher than the other two, and the increase for scientometrics was at an intermediate level. Based on this systematic analysis, the authors conclude that bibliometrics can be used as a general term for scientometrics and informetrics. In terms of the subject structure, bibliometrics attaches greater importance to the development of the three metrics and the application of bibliometric methods in scientific research management. On the other hand, scientometrics emphasises the quality of scientific research output and scientometric development trends.

Introduced by Blackert and Siegal (1979) and Nacke (1979) (cited in Egghe, 2005: 1312), informetrics became popular in the late 1980s with the organisation of international informetrics conferences (Egghe, 2005). Informetrics' focus is on the quantitative aspects of information in any form, and not just the records of publication documents or bibliographies (Ingwersen and Björneborn, 2004). It looks at the information transactions that are taking place on the internet (De Bellis, 2009). It also deals with the media, statistical analysis of the scientific text and hypertext systems, information measures in electronic libraries and information production and retrieval (Glänzel, 2003). Informetrics was to designate a general subfield of information science that concerned the statistical analysis of communication processes occurring in science and scientific enterprise (Andrés, 2009). Bibliometrics and scientometrics are intersecting but non-overlapping subsets of informetrics (Gingras, 2016). Owing to this character of informetrics, it is also treated as scientometric or webometric research (Rousseau et al., 2018). Informetrics is also an extension of bibliometrics to cover non-scholarly communities in which information is produced, communicated and used (Ingwersen and Christensen, 1997).

New metrics

The latest entrants in the genre are webmetrics, or webometrics, netometrics and cybermetrics. These metrics rely on the information available on the internet. They deal with the quantitative aspects of information to find out how information is generated, organised and disseminated (Björneborn and Ingwersen, 2004). They adopt the methods employed in bibliometrics, informetrics or scientometrics for the analysis of the information obtained from the internet resources.

Webometrics and cybermetrics, synonymously used, undertake studies of the quantitative aspects of information sources, structures and technologies on the World Wide Web using bibliometrics and informetrics approaches (Björneborn and Ingwersen, 2004). Webometrics, an analysis of electronic sources of information, covers a range of areas such as link analysis (study of hyperlinks between web pages), web citation analysis, search engine evaluation and descriptive studies of the web (Thelwall, 2008). Link counting, web mention counting and URL (Uniform Resource Locator) citation counts (Thelwall and Sud, 2011) are employed mainly in webometrics. In view of the practical challenges regarding measurements, collection and cleaning of data, webometrics cannot be used for research evaluation in the HSS (Archambault and Gagné, 2004).

There are other forms that cannot be omitted. Altmetrics, or alternate metrics, is used where the existing methods are inadequate. It consists of techniques that measure new forms of performing, discussing and communicating science through social media platforms (Rousseau et al., 2018) such as Twitter, Mendeley and Facebook. As a result of the engagement of social media, publications can be tracked instantly in real time by the count of page views, downloads, reads, shares, tweets, recommendations, saves, comments and reviews (Bornmann and Leydesdorff, 2014; Prathap, 2018). Altmetrics captures different forms of engagement relating to a paper, scientist or theory (Rousseau et al., 2018) and provides current data. True to its definition, altmetrics is based on the sources of scholarship on the internet. Alternative indicators have emerged as a standard part of scholarly communication infrastructure, and serve well in text-mining and impact agendas (Thelwall, 2019).

In brief, scientometrics is often used as bibliometrics or vice versa. The distinction between them is, however, thin due to their overlapping applications. While both have overlapping features and properties, scientometrics is meant for a more detailed study of the communication of scientific knowledge. Bibliometrics, on the other hand, deals with more general information about bibliographic records of publications. Many scholars do not make a clear-cut distinction between these two but use them interchangeably in their work. Glänzel (2003), for instance, treats bibliometrics and scientometrics synonymously. While referring to their works in this book, whether it is about bibliometrics or scientometrics, scientometrics is used, unless otherwise warranted, for consistency and to avoid ambiguity.

Theoretical foundations

The theoretical underpinnings of scientometrics render it a strong disciplinary foundation. This foundation is crucial for a methodology which is grounded in theories. Theoretical background provides a structure and a framework for research investigations and improves the validity of the knowledge that is produced. Derek John de Solla Price, who wanted to create a science of science on the basis of the quantitative analysis of scientific development, gave scientometrics the first theoretical basis it required (Gingras, 2016). Price proposed to analyse science as a collective phenomenon, following on the evolution of scientists and their publications (Gingras, 2016). Science, as he defines it, is what is published in journals, papers, reports and books and, in other words, that is embodied in the literature (Price, 1965).

Price's law of productivity

With his background in physics and history, Price was convinced that the structure of science can be studied by scientific methods (Price, 1965). Science of science, for him, was a new discipline initiated by historians, sociologists and economists who studied scientists. He found that the science of science has a stratum of knowledge explored in directions. The top stratum of administrators and experts on science policy examines new areas to find a base for their studies. The bottom stratum consists of those who do historical analyses and study the sociological and psychological characteristics of scientists, to build a picture of how science and scientists work. In between these two strata are the economists and statisticians (Price, 1965). Through these premises, Price developed his law of productivity.

The law of productivity helps to quantify research output and to make comparisons at levels that make sense. It can also assist in understanding the dynamics of the production of science and scientific knowledge. Many found Price's law of productivity worth applying to their studies (Barrios et al., 2013; Köseoglu et al., 2015, for instance, adopted it in their studies).

The science of science was essential as the substance of science in society has been growing and dominating. Price's quantitative approach showed the exponential growth of scientific knowledge. Of the two main strong streams of theoretical commitment in scientometrics, the quantitative history of science of Price is as inevitable as the normative sociology of science of Robert Merton (De Bellis, 2009). Merton's normative premises (norms such as universalism, communalism, disconnectedness, organised scepticism, originality, humility, integrity and intellectual honesty) are functional, as they facilitate the continuity of science as a large social activity (Patel, 1975).

Along with Price, the theoretical contributions of Alfred James Lotka, Samuel Bradford, George Kingsley Zipf and the like were vital in the evolution and development of scientometrics.

Lotka's law on frequency of publications

Alfred James Lotka (1840–1949), a mathematician, proposed a law of scientific productivity which formed the foundation of modern scientometrics (Ivancheva, 2008). Investigating the productivity patterns among scientists, Lotka (1926) discovered that only a fraction of researchers produces the majority of publications in their respective fields of inquiry. Inversely, it means a majority of researchers produce a minority of publications. If chemists, as they were his primary focus, are ranked according to the frequency of publication, then the number of chemists publishing n papers, $f(n)$, is more or less equal to a/n^2 (Chen and Leimkuhler, 1986). Lotka's law on the frequency of publication in any given subject field states that "the number of authors making n contributions is about $1/n^2$ of those making one; and the proportion of all contributors, that make a single contribution, is about 60 per cent" (Osareh and Mostafavi, 2011). In other words, out of all the authors in a given field, about 60 per cent publish only one paper, 15 per cent publish two and 7 per cent publish three (Osareh and Mostafavi, 2011). The law is called the "inverse square law", i.e., there is an inverse relation between the number of publications and the number of authors producing publications. It is the 20 per cent of the researchers who account for 80 per cent of the published documents; and 80 per cent of the researchers account for the 20 per cent of the published documents (Sugimoto and Larivière, 2018). The law has been applied not only in the natural sciences (Gupta, 1987), but also in the humanities (Köseoglu et al., 2015; Murphy, 1973; Nicholls, 1986; Tsai, 2015; Voos, 1974), and has produced insightful results about productivity.

The law of bibliographic scattering – Bradford

A British mathematician and librarian, Samuel C. Bradford (1878–1948), is credited with the law on the distribution of publications in a subject. Called the law of bibliographic scattering, it explains how the literature on a subject is scattered and distributed in journals. Examining the journals and publications in geophysics, Bradford discerned an inverse relationship between the number of papers published in any given subject area, and the number of journals in which these papers appeared. The bulk of the papers on a special subject would be published in a few journals that are devoted to the subject, or to the major subject of which the specific subject is a part, along with some general and borderline journals (Bradford, 1985 [1934]).

To Bradford, the law of distribution of papers on a given subject in scientific periodicals can be stated as: when scientific journals are arranged in the order of decreasing productivity of articles, they may be divided into a nucleus of periodicals more devoted to the subject and groups of zones, containing the same number of articles as the nucleus (Bradford, 1985 [1934]).

When journals carrying papers in a subject area are ranked in the decreasing order of productivity, and the number of papers contributed by each of the journals are counted, a core or nucleus of a few journals accounting for most of the papers in the subject area is known (Alabi, 1979; De Bellis, 2009). As a consequence of this, the majority of the citations are received by only relatively few journals, and a majority of the journals receive relatively few of the overall citations (Sugimoto and Larivière, 2018). Bradford called this the nucleus, which is the first zone of periodicals devoted to the given subject, which may also contain an article of interest meant for any other subject (Alabi, 1979). The remaining papers are spread over other journals.

According to the law, the distribution (scattering) of documents on a given subject follows a certain mathematical function, so that a growth of papers on a subject requires a growth in the number of journals/information sources (Hjørland and Nicolaisen, 2005). Bradford's law provides mechanisms to select the most productive and relevant journals that cover a subject area (Alvarado, 2016). The law derives from the basic unity of science that every scientific field is related to every other field (Garfield, 1980). In citation and impact studies the meaningful purposes of the law are evident. Both scholars and institutions alike find its application valuable for informed decisions pertaining to research, publication outlets and impact. Adopting the law of Bradford, many scientometric studies have appeared and continue to appear in the literature. The application of this theory has found its way to the study of disciplines in the HSS as well.[5]

The law of word frequency – Zipf

The basis of the law of word frequency is the frequency of words in any given publication. George Kingsley Zipf (1902–1950), a linguist and philosopher, examined the frequency of words to develop his law. Zipf realised the utility of studying speech as a natural phenomenon and a peculiar form of behaviour that has communicative gestures. His ideas were first introduced in his two books, *The Psycho-Biology of Language* (1935) and *Human Behaviour and the Principle of Least Effort* (1949). The principle of least effort, according to Zipf (1949), means that a person will strive to solve their problems by minimising the total work that they must expend in solving both their immediate and future problems.

For Zipf, language is a complex tool of behaviour and its structure cannot be separated from the personal, social and cultural backgrounds of the speakers (De Bellis, 2009). In other words, language reflects one's views, vision, attitude, approach and philosophy. The law of Zipf, which became a framework for several studies in the area, states that if the number of different words occurring once in a given sample is taken as x, for the number of different words occurring twice, thrice, four times or n times in the sample, one can find the progression according to the inverse square which is valid for over 95 per cent of all different words used in the sample (Zipf, 1935).

Further, Zipf's law states that "if one takes the words making up an extended body of text and ranks them by frequency of occurrence, then the rank r multiplied by its frequency of occurrence g (r) will be approximately constant" (Chen and Leimkuhler, 1986: 308). Zipf's formula is $ab^2=k$ where a is the number of words occurring b times. He reiterated that the $ab^2=k$ relationship is valid only for the less frequently occurring words, which represent the most part of the vocabulary in use and not always a great majority of occurrences (Zipf, 1935). Zipf was attempting to seek the relationship between rank and the frequency of linguistic and social units (Powers, 1998). Applications of Zipf's law are evident in the HSS (Saxena et al., 2003, for instance). Zipf's law has made a substantial contribution to the field of linguistics and semantics. It has also been benefitted by the modifications proposed by others (Powers, 1998).

Zipf's law, like Pareto's distribution, can be treated as power law. When the probability of measuring a quantity, productivity for instance, varies inversely as a power of that value, it can be explained by power law (Newman, 2005). Power law is used in both natural and social sciences, either in the study of populations or citations. Newman (2005) shows the variety of applications of power law from the study of earthquakes to the sale of books.

The Matthew Effect – Merton

The Matthew Effect of Robert Merton (1910–2003) provided a different theoretical basis for the sociology of science. Merton explained how the recognition for scientific discoveries is given, which is likely to be allocated to those who already have received recognition in the past. It is "the man who's best known gets more credit, an inordinate amount of credit" (Merton, 1988: 607). The Matthew Effect is about the misallocation of recognition for scientific work. He drew this principle from the biblical parable in the New Testament (Matthew, 13:12 and 25:59), which states: "For unto everyone that hath shall be given, and he shall have abundance; but from that hath not shall be taken away even that which he hath".

The Matthew Effect is "the accruing of the large increments of peer recognition to scientists of great repute for their particular contributions as against to the minimizing or withholding of such recognition for scientists who have not yet made their mark" (Merton, 1988: 609). Merton argued that, in the case of independent multiple discoveries or collaboration in a scientific enterprise, the more eminent of the two or more collaborators will get the lion's share of the credit, even if that person did only a small amount of work in the said discovery or collaboration or publication.

The Matthew Effect has applications and relevance in the scientometric study of citations and impact, and in areas of rewards, recognition and benefits. It has also been applied to study the growth and structure of scientific information (Sugimoto and Larivière, 2018). The Matthew Effect could be generalised to all scientific fields by hypothesising that characteristics other than the scientific work itself influence the evaluation of a new scientific work (Cole, 2004). Stephen Cole, a student, associate and later a critique of Merton, believes that the Matthew Effect is a nice sociological theory which shows how the underdogs are taken advantage of in the system and who are not given due credit. It also shows how the social characteristic of eminence is more important as the cognitive content in the evaluation of new work (Cole, 2004).

Both Jonathan Cole and Stephen Cole (1973) in their well-known book, *Social Stratification in Science*, discuss the same social inequalities in the scientific community. This is done by examining the factors that determine the allocation of recognition and honours to scientists and their departments. Science is viewed as a stratified social institution based on universalistic criteria. In this system, individuals may be displaced if they are inactive or their accomplishments are not significant. This has relevance in the study of some important aspects of the scientific system. Nevertheless, in the current scheme of things in which citation count is a criterion, all partners are able to get the same credit for their publications. If a publication is authored by two scholars and it earns 100 citations in its life, each of the author gets the credit of 100 citations and not a share of 50 per cent. At the same time, citations do not take into account the standing and reputation of authors, but these definitely influence their citations.

Delayed recognition is a related concept here. Sometimes, publications receive little recognition in their beginning years, but may receive increased attention and recognition in later years. Delayed recognition is referred to as Mendel's syndrome, named after the founder of genetics, Gregor Mendel (1822–1884). His discoveries in plant genetics remained unused for 34 years. In plants, there are late bloomers or those which bloom ahead of their time. In scientometrics, delayed recognition is known as sleeping beauty, a term introduced by Anthony van Raan in 2004 (Rousseau, 2018). Sleeping beauty is a publication that goes unnoticed or sleeps for a long time since its appearance and then suddenly attracts lot of attention, as though awakened by a prince (van Raan, 2004). Although delayed recognition of publications is not a clear-cut phenomenon, it can be examined by using citation data, as done by Rousseau (2018) in his study.

Theory of citation – Small

Scientometricians were keen to adopt theories from other sources in their analyses. For instance, theories from the semiotic tradition, the study of signs and symbols, have been incorporated to study scientific communication (Sugimoto and Larivière, 2018). The canonical theory of "cited documents as concept symbols" belongs to this tradition (Sugimoto and Larivière, 2018). Small's exposition, described in detail in the paper written in 1978 under the same title, is to make sense of why authors cite the work of others. This was then not investigated adequately. According to Small (1978), not much attention has been paid to the understanding of an author's motivation for citing a particular work. He regarded citations as symbols of concepts of ideas, as cited works embody ideas that authors discuss in their work. For him, an object stands for an idea and, for citations, the cited document is actually the object and the idea is stated in the text, which cites the document. Small regarded ideas or concepts as those residing in the mind and expressed in language. References are sources that give further meaning to the text (Small, 1978). He applied the terms to abstract, theoretical formulations, experimental findings, methodologies, types of data and metaphysical notions. By putting forward this theory, Small was challenging scholars to go further from the statistical analysis of citations, and to understand scientific communication at deeper levels of analysis and insights.

As elaborated above, substantial theoretical contributions have been made by a range of scholars, from Derek de Solla Price to Robert Merton, to Jonathan Cole and Stephen Cole, which have provided the required support to lay the theoretical foundations for scientometric studies. Garfield's lifelong work in the field initiated several developments in scientometrics, with his citation analysis still being one of the strong areas of scientometric studies. Theories serve different purposes in scientometrics as they have different foci, such as words, productivity, language, recognition, citations and references. The complexities of science and its production are disentangled with the appropriate application of these theories.

Scientometric paradigms

The paradigms of scientometrics have not received due attention in the scientometric literature, but knowledge about scientometric paradigms is necessary to understand more about scientometrics and its methods. Some methodological discussion on the methods used in scientometrics will clarify its uses. Leydesdorff (2001b), in *The Challenge of Scientometrics*, offers a detailed elaboration on methodological issues that are pertinent to the study of science. Prescribing certain requirements for the methods, he thinks that methods should allow for the import of data and results from other types of studies; and data analysis should support the translation among various paradigms that are being used in science. Meanwhile, methods should allow for variations in the types of theories and methods that use the same or similar data, which he calls the requirement of multiple paradigms. In the aggregation and decomposition of data, methods should allow for the control of the relations among the levels of aggregation and in the measurement scale of observations. The requirement of multivariate statistics is essential for the methods in science. In short, one should be able to evaluate relations among variables at different levels of aggregation, reconstruct scientific developments from mere occurrences of the nominal variables, and make predictions about new events (Leydesdorff, 2001a). This is what scientometrics is able to achieve in its field.

More elaboration on scientometric paradigms is provided in the papers, including those by Small (2003) and Morris (2005). They deliberate these from the definition of Thomas Kuhn, whose well-cited work, *The Structure of Scientific Revolutions* (1970), deals with the paradigms for normal science. To him, paradigms are used by a group of researchers to define the problems and the methods to resolve them. These are actual practices in law, theory, application and instrumentation that provide models from which spring particular coherent traditions of scientific research. They are the accepted models or patterns. Scholars whose research is based on shared paradigms are to be committed to the same rules and standards for the common scientific practice.

In a previous work, prior to the above cited monograph published in 1970, Kuhn (1963) briefly explained paradigm as a fundamental scientific achievement, which includes theory and applications to the results of experiment and observation. It is an accepted achievement as a group and its members attempt to extend and exploit it in a variety of ways. Sometimes, problems cannot be solved within the confines of a paradigm, causing a paradigm shift, which changes the fundamental ways of thinking about a subject or discipline. In the event of a paradigm shift or a new paradigm, a new set of puzzles is to be solved and a new set of methods needs to be found to solve these puzzles. However, there are also some who would continue to cling to one or other of the older views. Kuhn shows how the emergence of a paradigm affects the structure of the group that practises the field. The new paradigm implies a new and more rigid definition of the field.

Kuhnian aspects of a paradigm can support an understanding and explanation about the paradigms used in scientometrics. A scientific discipline in its stages of development, as Barnes (1969) observed, has a single paradigm which will become accepted throughout the field. Although not very big in size, the community of scientometricians and the field of scientometrics follow certain approaches to their study of problems and the methods they employ to resolve them. The practice in the community, in terms of law, theory, application and instrumentation, is a shared one.

Further, the sociology of science is fractured into two research areas, namely scientometrics, and science and technology studies (Sugimoto and Larivière, 2018). These branches have their own paradigmatic approaches, positivist and constructivist, and quantitative and qualitative. Often one is pitted against the other for their methodological preferences, strategies, approaches and advantages. Sugimoto and Larivière (2018) argue that contemporary approaches to the science of science should take both paradigmatic approaches into account, triangulating theories and methods. They believe that such an approach is complementary rather than adversarial, and lends a more robust lens to understand science and knowledge.

Cotemporary scientometrics follows positivist and post-positivist paradigms. The methods adopted in the early years have been purely quantitative. Scientometric practitioners then began to adopt a mixed method approach in dealing with scientometric problems. They find value and significance in applying both quantitative and qualitative methods. As will be shown in the following chapters, the application of relevant paradigms, not rigidly restricted to one or the other, has a positive effect on the development and future of scientometric studies. For the study of the HSS, a mixed approach will be more rewarding than a single approach. While showing the complementarity of both quantitative and qualitative methods, the importance of qualitative paradigms can only be undermined at the expense of the neglect of the untapped qualitative data. In the next chapter the uses and applications of scientometrics are elaborated.

Notes

1 The original Polish version, "Nauka o nauce", appeared in 1935 in *Nauka Polska* (vol. 20, pp.1–12) (Zdrenka, 2006).
2 Apart from these basic publication types, some databases have documents such as abstracts, exhibit and book reviews, dance performance reviews, editorials, hardware and software reviews, letters, performance reviews, music scores, news items, poetry and TV, radio and theatre reviews.
3 This enlarged edition, *Little Science, Big Science … and Beyond,* appeared in 1986 with additional chapters.
4 The purpose of this application is to analyse the potential interest of researchers in the method and how it can be used for research policy purposes (Torres-Salinas and Robinson-García, 2016).
5 Humanities disciplines are archaeology, arts, history, language, literature, media studies, philology, philosophy, musicology and theology. Social sciences, by definition, include any discipline that deals with human behaviour in social situations. They cover a broad range of disciplines such as demography and social statistics, development studies, human geography, environmental planning, economics, management and business studies, education, social anthropology and linguistics, law, economy and social history, political and

international relations, psychology, sociology, science and technology studies, social policy and social work (Economic and Social Research Council, https://esrc.ukri.org/about-us/what-is-social-science/social-science-disciplines/).

References

Alabi, G. (1979). Bradford's law and its application. *International Library Review, 11*, 151–158. https://doi.org/10.1016/0020-7837(79)90044-X

Alvarado, R. U. (2016). Growth of literature on Bradford's Law. *Investigación Bibliotecológica, 30*, 51–72. https://doi.org/10.1016/j.ibbai.2016.06.003

Andrés, A. (2009). *Measuring Academic Research: How to Undertake a Bibliometric Study* Oxford: Chandos Publishing.

Archambault, É. & Gagné, É.V. (2004). *The Use of Bibliometrics in the Social Sciences and Humanities.* Montreal: Social Sciences and Humanities Research Council of Canada (SSHRCC). www.science-metrix.com/pdf/SM_2004_008_SSHRC_Bibliometrics_Social_Science.pdf

Ball, N. T. (1950). The Royal Society Scientific Information Conference, 21 June–2 July 1948: Report and papers submitted. *The Library Quarterly: Information, Community, Policy, 20*, 45–47. www.ncbi.nlm.nih.gov/pmc/articles/PMC194801/

Barnes, S. B. (1969). Paradigms – scientific and social. *Man, New Series, 4*, 94–102. DOI: 10.2307/2799267

Barrios, M., González-Teruel, A., Cosculluela, A., Fornieles, A., Ortega, L. & Turbany, J. (2013). Temporal evolution, structural features and impact of standard articles and proceedings papers: A case study in blended learning. In J. Gorraiz, E. Schiebel, C. Gumpenberger, M. Hörlesberger & H. Moed (Eds), *14th International Society of Scientometrics and Informetrics Conference* (Vol. 2, pp. 2156–2158). Vienna: AIT Austrian Institute of Technology GmbH.

Belter, C. W. (2015). Bibliometric indicators: Opportunities and limits. *Journal of Medical Library Association, 103*, 219–221. DOI: 10.3163/1536-5050.103.4.014

Bernal, J. D. (1939). *The Social Function of Science.* London: George Routledge and Sons, Ltd.

Björneborn, L. & Ingwersen, P. (2004). Toward a basic framework for webometrics. *Journal of the American Society for Information Science and Technology, 55*, 1216–1227. https://doi.org/10.1002/asi.20077

Blackert, L. & Siegel, S. (1979). Ist in der wissenschaftlich-technischen Information Platz für die Informetrie? *Wissenschaftliches Zeitschrift TH Ilmenau, 25*, 187–199.

Bornmann, L. (2014). Is there currently a scientific revolution in scientometrics? Letter to the Editor. *Journal of the Association for Information Science and Technology, 65*, 647–648. https://doi.org/10.1002/asi.23073

Bornmann, L. & Leydesdorff, L. (2014). Scientometrics in a changing research landscape. *EMBO Reports, 15*, 1228–1232. DOI: 10.15252/embr.201439608

Bradford, S. C. (1985[1934]). Sources of information on specific subjects. *Journal of Information Science, 10*, 173–180. https://doi.org/10.1177/016555158501000407

Broadus, R.N. (1987). Early approaches to bibliometrics. *Journal of the American Society for Information Science and Technology, 38*, 127–129. https://doi.org/10.1002/(SICI)1097-4571(198703)38:2<127::AID-ASI6>3.0.CO;2-K

Brookes, B. C. (1988). Comments on the scope of bibliometrics. In L. Egghe & R. Rousseau (Eds), *Informetrics 87/88: Select Proceedings of the First International Conference on Bibliometrics and Theoretical Aspects of Information Retrieval, Diepenbeek, Belgium, 25–28 August 1987* (pp. 29–63). Amsterdam: Elsevier Science Publishers.

Cantú-Ortiz, F. J. (2018). Data analytics and scientometrics: The emergence of research analytics. In F. J. Cantú-Ortiz (Ed.), *Research Analytics: Boosting University Productivity and Competitiveness through Scientometrics* (pp. 1–11). Boca Raton, FL: CRC Press.

Chen, Y. H. & Leimkuhler, F. F. (1986). A relationship between Lotka's Law, Bradford's Law, and Zipf's Law. *Journal of the American Society for Information Science, 37*, 307–314. https://doi.org/10.1002/(SICI)1097–4571(198609)37:5<307::AID-ASI5>3.0.CO;2–8

Cole, J. R. & Cole, S. (1973). *Social Stratification in Science*. Chicago, IL: The University of Chicago Press.

Cole, S. (2004). Merton's contribution to the sociology of science. *Social Studies of Science, 34*, 829–844. https://doi.org/10.1177/0306312704048600

De Bellis, N. (2009). *Bibliometrics and Citation Analysis: From the Science Citation Index to Cybermetrics*. Lanham, MD: The Scarecrow Press, Inc.

Egghe, L. (2005). Editorial. Expansion of the field of informetrics: Origins and consequences. *Information Processing & Management, 41*, 1311–1316. DOI: 10.1016/j.ipm.2005.03.011

Elkana, Y., Lederberg, J., Merton, R. K., Thackray, A. & Zuckerman, H. (Eds) (1974). *Toward a Metric of Science: The Advent of Science Indicators*. New York: John Wiley & Sons.

Garfield, E. (1955). Citation indexes for science: A new dimension in documentation through association of ideas. *Science, 122*, 108–111. DOI: 10.1126/science.122.3159.108

Garfield, E. (1975). The Social Sciences Citation Index, more than a tool. *Current Contents, 12*, 6–9. www.garfield.library.upenn.edu/essays/v2p241y1974-76.pdf

Garfield, E. (1980). Bradford's law and related statistical patterns. *Current Contents, 19*, 5–12. www.garfield.library.upenn.edu/essays/v4p476y1979-80.pdf

Garfield, E. (1985). George Sarton: The father of the history of science. Part 1. Sarton's early life in Belgium. *Current Contents, 25*, 3–9. www.garfield.library.upenn.edu/essays/v8p248y1985.pdf

Garfield, E. (1992). Science historian I. B. Cohen reviews social studies of science by sociologist Bernard Barber. *Current Contents, 9*, 3–9. www.garfield.library.upenn.edu/essays/v15p028y1992-93.pdf

Garfield, E. (2007). From the science of science to scientometrics. Visualizing the history of science with Histcite software. In D. Torres-Salinas & H. F. Moed (Eds), *11th International Conference of the International Society for Scientometrics and Informetrics* (Vol. 1, pp. 21–26). Madrid: Centre for Scientific Information and Documentation (CINDOC) of the Spanish Research. https://doi.org/10.1016/j.joi.2009.03.009

Gingras, Y. (2016). *Bibliometrics and Research Evaluation: Uses and Abuses*. Cambridge, MA: MIT Press.

Glänzel, W. (2003). Bibliometrics as a research field: A course on theory and application of bibliometric indicators. Retrieved from http://yunus.hacettepe.edu.tr/~tonta/courses/spring2011/bby704/bibliometrics-as-a-research-field-Bib_Module_KUL.pdf, 21 May 2019.

Glänzel, W. & Schoepflin, U. (1994). Little scientometrics, big scientometrics … and beyond? *Scientometrics, 30*, 375–384. https://doi.org/10.1007/BF02018107

Godin, B. (2006). On the origins of bibliometrics. *Scientometrics, 68*, 109–133. DOI: 10.1007/s11192-006-0086-0

Godin, B. (2007). From eugenics to scientometrics: Galton, Cattell, and men of science. *Social Studies of Science, 37*, 691–728. https://doi.org/10.1177/0306312706075338

Gross, P. L. K. & Gross, E. M. (1927). College libraries and chemical education. *Science, 66*, 385–389. DOI: 10.1126/science.66.1713.385

Gupta, D. K. (1987). Lotka's law and productivity of entomological research in Nigeria for the period 1900–1973. *Scientometrics, 12*, 33–46. https://doi.org/10.1007/BF02016688

Hertzel, D. H. (2003). Bibliometrics history. In M. A. Drake (Ed.), *Encyclopedia of Library and Information Science* (Second ed., Vol. 1, pp. 288–328). New York: Marcel Dekker, Inc.

Hjørland, B. & Nicolaisen, J. (2005). Bradford's law of scattering: Ambiguities in the concept of "subject". In F. Crestani & L. Ruthven (Eds), *5th International Conference on Conceptions of Library and Information Sciences* (pp. 96–106). Glasgow: Springer-Verlag.

Ingwersen, P. & Björneborn, L. (2004). Methodological issues of webmetric studies. In H. F. Moed, W. Glänzel & U. Schmoch (Eds), *Handbook of Quantitative Science and Technology Research: The Use of Publication and Patent Statistics in Studies of S&T Systems* (pp. 339–369). Dordrecht, The Netherlands: Kluwer. DOI: 10.1007/1-4020-2755-9_16

Ingwersen, P. & Christensen, F. H. (1997). Data set isolation for bibliometric online analyses of research publications: Fundamental methodological issues. *Journal of the American Society for Information Science, 48*, 205–217. https://doi.org/10.1002/(SICI)1097–4571(199703)48:3<205::AID-ASI3>3.0.CO;2-0

Ivancheva, L. (2008). Scientometrics today: A methodological overview. *Collnet Journal of Scientometrics and Information Management Decision, 2*, 47–56. https://doi.org/10.1080/09737766.2008.10700853

Köseoglu, M. A., Sehitogluc, Y. & Parnell, J. A. (2015). A bibliometric analysis of scholarly work in leading tourism and hospitality journals: The case of Turkey. *Anatolia – An International Journal of Tourism and Hospitality Research, 26*, 359–371. https://doi.org/10.1080/13032917.2014.963631

Kuhn, T. S. (1963). The function of dogma in scientific research. In A. C. Crombie (Ed.), *Scientific Change* (pp. 347–369). London: Heinermann.

Kuhn, T. S. (1970). *The Structure of Scientific Revolutions* (Second ed.). Chicago, IL: University of Chicago Press.

Lazarsfeld, P. F. (1961). Notes on the history of quantification in sociology – Trends, sources and problems. *Isis, 52*, 277–333. DOI: 10.1086/349473

Leydesdorff, L. (2001a). Scientometrics and science studies. *BMS: Bulletin of Sociological Methodology, 71*, 79–91. https://doi.org/10.1177/075910630107100105

Leydesdorff, L. (2001b). *The Challenge of Scientometrics: The Development, Measurement, and Self-Organization of Scientific Communities* (Second ed.). Irvine, CA: Universal Publishers. DOI: 10.2139/ssrn.3512486

Lotka, A. J. (1926). The frequency of distribution of scientific productivity. *Journal of the Washington Academy of Science, 16*, 317–323.

Merton, R. K. (1988). The Matthew Effect in science, II: Cumulative advantage and the symbolism of intellectual property. *Isis, 79*, 606–623. DOI: 10.1086/354848

Morris, S. A. (2005). Manifestation of emerging specialties in journal literature: A growth model of papers, references, exemplars, bibliographic coupling, cocitation, and clustering coefficient distribution. *Journal of the American Society for Information Science and Technology, 56*, 1250–1273. https://doi.org/10.1002/asi.20208

Murphy, L. J. (1973). Lotka's law in the humanities? *Journal of the American Society for Information Science and Technology, 24*, 461–462. https://doi.org/10.1002/asi.4630240607

Nacke, O. (1979). Informetrie: eine neuer Name für eine neue Disziplin. *Nachrichten für Documentation, 30*, 219–226.

Nalimov, V. V. & Mulchenko, Z. M. (1969). *Naukometriya. Izuchenie nauki kak informatsionnogo protsessa (Scientometrics. Study of Science as an Information Process)*. Moscow: Nauka. English version: *Measurement of Science. Study of the Development of Science as an Information Process.* Washington, DC: Foreign Technology Division, US Air Force Systems Command. Retrieved from http://garfield.library.upenn.edu/nalimovmeasurementofscience/chapter1.pdf. 20 March 2019.

Newman, M. E. J. (2005). Power laws, Pareto distributions and Zipf's law. *Contemporary Physics, 46*, 323–351. https://doi.org/10.1080/00107510500052444

Nicholls, P. T. (1986). Empirical validation of Lotka's law. *Information Processing and Management, 22*, 417–419. https://doi.org/10.1016/0306-4573(86)90076-2

Osareh, F. & Mostafavi, E. (2011). Lotka's law and authorship distribution in computer science using Web of Science (WoS) during 1986–2009. *Collnet Journal of Scientometrics and Information Management, 5*, 171–183. https://doi.org/10.1080/09737766.2011.10700911

Ossowska, M. & Ossowski, S. (1964). The science of science. *Minerva, 3,* 72–82. https://doi.org/10.1007/BF01630150

Patel, P. J. (1975). Robert Merton's formulations in sociology of science. *Sociological Bulletin, 24,* 55–75. https://doi.org/10.1177/0038022919750104

Peritz, B. C. (1988). Bibliometric literature: A quantitative analysis. In L. Egghe & R. Rousseau (Eds), *Informetrics 87/88: Select Proceedings of the First International Conference on Bibliometrics and Theoretical Aspects of Information Retrieval, Diepenbeek, Belgium, 25–28 August 1987* (pp. 165–173). Amsterdam: Elsevier Science Publishers.

Powers, D. M. W. (1998). *Applications and Explanations of Zipf's Law.* Paper presented at the Joint Conferences on New Methods in Language Processing and Computational Natural Language Learning, Sydney, Australia, 11–17 January. DOI: 10.3115/1603899.1603924

Prathap, G. (2018). Eugene Garfield: From the metrics of science to the science of metrics. *Scientometrics, 114,* 637–650. https://doi.org/10.1007/s11192-017-2525-5

Price, D. J. d. S. (1961[1975]). *Science Since Babylon* (Enlarged ed.). New Haven and London: Yale University Press.

Price, D. J. d. S. (1963). *Little Science, Big Science.* New York: Columbia University Press.

Price, D. J. d. S. (1965). The science of science. *Bulletin of the Atomic Scientists, 21,* 2–8.

Price, D. J. d. S. (1986). *Little Science, Big Science … and Beyond.* New York: Columbia University Press.

Riecken, H. W. (1980). Vital signs for basic research in the behavioral and social sciences. *Scientometrics, 2,* 435–437. https://doi.org/10.1007/BF02095087

Rousseau, R. (2018). Delayed recognition: Recent developments and a proposal to study this phenomenon as a fuzzy concept. *Journal of Data and Information Sciences, 3,* 1–13. DOI: 10.2478/jdis-2018-0011

Rousseau, R., Egghe, L. & Guns, R. (2018). *Becoming Metric-Wise: A Bibliometric Guide for Researchers.* Cambridge, MA: Chandos Publishing.

Rushton, J. P. & Meltzer, S. (1981). Research productivity, university revenue, and scholarly impact (citations) of 169 British, Canadian and United States universities (1977). *Scientometrics, 3,* 275–303. https://doi.org/10.1007/BF02021122

Sarton, G. (1927). *Introduction to the History of Science: From Homer to Omar Khayyam* (Vol. 1). Baltimore, MD: The Williams and Wilkins Co., for the Carnegie Institution of Washington.

Sarton, G. (1931). *Introduction to the History of Science: From Rabbi ben Ezra to Roger Bacon* (Vol. 2, Parts 1 and 2). Baltimore, MD: The Williams and Wilkins Co., for the Carnegie Institution of Washington.

Sarton, G. (1936). *The Study of the History of Science.* Harvard, MA: Harvard University Press.

Sarton, G. (1947). *Introduction to the History of Science: Science and Learning in the Fourteenth Century* (Vol. 3, Parts 1 and 2). Baltimore, MD: The Williams and Wilkins Co., for the Carnegie Institution of Washington.

Sarton, G. (1952). *History of Science* (Vol. 1). Harvard, MA: Harvard University Press.

Sarton, G. (1959). *History of Science: Hellenistic Science and Culture in the Last Three Centuries BC* (Vol. 2). Harvard, MA: Harvard University Press.

Saxena, A., Jauhari, M. & Gupta, B. M. (2003). Zipf's law in a random text from English with a new ranking method. In G. Jiang, R. Rousseau & Y. Wu (Eds), *Proceedings of the 9th International Conference on Scientometrics and Informetrics* (pp. 271–279). Dalian, China: Dalian University of Technology Press.

Siluo, Y. & Qingli, Y. (2017). *Are Scientometrics, Informetrics, and Bibliometrics Different? Conference Proceedings of the 16th International Conference on Scientometrics and Informetrics* (pp. 1507–1518). Wuhan, China.

Sivertsen, G. (2016). Publication-based funding: The Norwegian model. In M. Ochsner, S. E. Hug & H.-D. Daniel (Eds), *Research Assessment in the Humanities Towards Criteria and Procedures* (pp. 79–90). Zürich: Springer. DOI: 10.1007/978-3-319-29016-4_7

Small, H. (1978). Cited documents as concept symbols. *Social Studies of Science, 8*, 327–340. https://doi.org/10.1177/030631277800800305

Small, H. (2003). Paradigms, citations, and maps of science: A personal history. *Journal of the American Society for Information Science and Technology, 54*, 394–399. https://doi.org/10.1002/asi.10225

Small, H. G. & Crane, D. (1979). Specialties and disciplines in science and social science: An examination of their structure using citation indexes. *Scientometrics, 1*, 445–461. https://doi.org/10.1007/BF02016661

Sugimoto, C. R. & Larivière, V. (2018). *Measuring Research: What Everyone Needs to Know*. New York: Oxford University Press.

Szántó-Várnagy, À., Pollner, P., Vicsek, T. & Farkas, I. J. (2014). Scientometrics: Untangling the topics. *National Science Review, 1*, 343–345. https://doi.org/10.1093/nsr/nwu027

Thelwall, M. (2008). Bibliometrics to webometrics. *Journal of Information Science, 34*, 605–621. DOI: 10.1177/0165551507087238

Thelwall, M. (2019). New developments in scientometric and informetric research. Keynote lecture. In G. Catalano, C. Daraio, M. Gregori, H. F. Moed & G. Ruocco (Eds), *Proceedings of the 17th Conference of the International Society for Scientometrics and Informetrics* (Vol. I, pp. xxxvii–xxxviii). Rome: Edizioni Efesto.

Thelwall, M. & Sud, P. (2011). A comparison of methods for collecting web citation data for academic organizations. *Journal of the American Society for Information Science and Technology, 62*, 1488–1497. https://doi.org/10.1002/asi.21571

Torres-Salinas, D. & Robinson-García, N. (2016). The time for bibliometric applications. Letter to the Editor. *Journal of the American Society for Information Science and Technology, 67*, 1014–1015. https://doi.org/10.1002/asi.23604

Tsai, H.-H. (2015). The research trends forecasted by bibliometric methodology: A case study in e-commerce from 1996 to July 2015. *Scientometrics, 105*, 1079–1089. https://doi.org/10.1007/s11192-015-1719-y

van Raan, A. F. J. (2004). Sleeping Beauties in science. *Scientometrics, 59*, 461–466. DOI: 10.1023/B:SCIE.0000018543.82441.f1

Vinkler, P. (2010). *The Evaluation of Research by Scientometric Indicators*. Oxford: Chandos Publishing.

Voos, H. (1974). Lotka and information science. *Journal of the American Society for Information Science. Brief Communication, 25*, 270–272. https://doi.org/10.1002/asi.4630250410

Yitzhaki, M. (1988). The language barrier in the humanities: Measures of language self-citation and self-derivation – The case of biblical studies. In L. Egghe & R. Rousseau (Eds), *Informetrics 87/88: Select Proceedings of the First International Conference on Bibliometrics and Theoretical Aspects of Information Retrieval, Diepenbeek, Belgium, 25–28 August 1987* (pp. 301–314). Amsterdam: Elsevier Science Publishers.

Zdrenka, M. T. (2006). Maria Ossowska: Moral philosopher or sociologist of morals? *Journal of Classical Sociology, 6*, 311–331. https://doi.org/10.1177/1468795X06069681

Zhang, L., Thijs, B. & Glänzel, W. (2013). What does scientometrics share with other "Metrics" Sciences? *Journal of the American Society for Information Science and Technology, 64*, 1515–1518. DOI: 10.1002/asi.22834

Zipf, G. K. (1935). *The Psycho-Biology of Language*. Cambridge, MA: MIT Press.

Zipf, G. K. (1949). *Human Behavior and the Principle of Least Effort: An Introduction to Human Ecology*. Cambridge, MA: Addison-Wesley Press.

2

APPLICATIONS AND USES OF SCIENTOMETRICS

Introduction

Scientometrics for the study of scientific fields, research and related matters is evident from the discussion in the previous chapter. The applications are quite broad. Pritchard and Witting (1981) note that scientometric data can be used as a visible sign of an underlying problem of social structure that relates to individual differences such as gender, promotion and creativity. The general growth and development of the social structure within a subject or discipline can be investigated by adopting a scientometric approach. In the evaluation of organisations, research sponsorship and government policies, scientometrics is an excellent research tool. In the evaluation of countries, comparisons between countries and their science policies, scientometrics is an essential tool. Pritchard and Witting assert that raw scientometric data is useful for operations research and mathematical models, and for the study of the distributions which are of interest to the social sciences.

Andrés (2009) lists three subareas for the applications of scientometrics: methodology research, scientific disciplines and science policy. Generally, scientometrics is adopted to the study of research performance, interdisciplinarity, collaboration, the structure of scientific fields and the impact of research (Anninos, 2014). The complexities of disciplines and subjects are made known through the study of publications, citations and their impact.

However, the uses and applications go beyond the study of scientific publications, their impact and citations. Scholars employ scientometrics for a range of applications that lead to a better understanding of science and the HSS. This chapter elaborates on the multifarious uses and applications of scientometrics and the use of scientometric data.

Uses and applications

Scientometrics is not an old field of study. Its predecessor and contemporary method of bibliometrics was coined and originated only in 1969.[1] This field of information science according to Robert Broadus (1987), who recollected the earliest approaches to the method, had a respectable history that promised developments in the field today.

Fundamentally, scientometrics is concerned with all quantitative aspects related to the production and dissemination of scientific knowledge (De Bellis, 2009). The set of instruments applied in scientometrics is observation, measurement, mathematical processing, comparison, classification and visualisation (Ivancheva, 2008). In addition to these, several other instruments, including advanced statistical procedures, exist.

Basics for applications

Scientometrics helps to gather a host of dimensions of scientific research, varying from the count of publications over the years to the impact of publications and to scientific collaboration. Documents constitute the basic unit of analysis for scientometric studies. Documents include, inter alia, journal articles, chapters, books, monographs, reports, theses, patents, notes, reviews and conference proceedings. Groups of scientists, institutions and countries (De Bellis, 2009) also make firm units of analysis. From these basic units, many more new variables or forms of variables can be created by splitting, merging or transforming. Indicators are developed from variables for measurement purposes, which take scientometrics to higher levels of analysis. Analysis can begin from a micro level and extend to macro levels (Wallin, 2005), from a single subject to several subjects or disciplines, or from a single country to several countries. The analysis of documents, for instance the documents in a subject area, can proceed with an individual, an institution or a single country. From this point, further complex analyses are performed for multiple cases of individuals, institutions and countries.

Glänzel (2003) identifies three levels of aggregation (micro, meso and macro) in scientometric research, which require their own methodological and technological approaches. At the micro level are the publication outputs of individuals and research groups. Publication outputs of institutions and studies of scientific journals are at the meso level. The macro level concerns publication outputs of regions, countries and supranational aggregations.

Scientometrics works on documents of publications that are found in databases, journal publications, books, proceedings and on the web. As these documents have not been processed or analysed, they constitute the primary sources of data for researchers. Owing to this nature of the data, it is a great attraction for students and researchers to undertake studies on data that has already been collected and stored. This is, however, not the case when scholars want to produce data of their own. They collect data directly from journal publications or from other sources. If an analysis of publications in a specific journal is to be carried out, researchers will have to manually collect the data from the chosen journal(s) and process them for analysis.

A range of techniques is employed in scientometrics. Analysis of word frequency, citation, authorship, co-citation, coauthor and co-word, and count of author, research groups, institutions or country (Benckendorff and Zehrer, 2013) are prominent among them.

Co-citation analysis

Co-citation, to take an example, is about the relation between citations. It is a technique to draw a comprehensive map of science (Leydesdorff, 1987). By citing two documents in one article, a co-citation link can be established. These links are counted for a year or so by using the Boolean AND in the search option (Leydesdorff, 1987). The relative citation rate is the ratio between the summation of observed values (the actual number of citations accrued to a publication) and a summation of expected values (average citation rate of the journal in which the paper is published) for all papers published by a country in a given research field. Citation analysis can be done for the total number of citations earned by papers of a scholar, institution, region, country or a group of countries for any chosen period of time. Studies of this kind have appeared in the HSS disciplines too (Santarem and Oliveira, 2009, for instance).

A relevant concept in citation, namely half-life, may be introduced here. The half-life citation is the number of retrospective years required to reach 50 per cent of the cited references (Garfield, 2007). In other words, cited half-life refers to the time span from the current year to the year in which accumulated number of citations has reached 50 per cent of the total citations (Tsay, 2015). Physical scientists use the term to denote decay, that is, the time required for the disintegration of half the atoms of a sample of a radioactive substance (Burton and Kebler, 1960). In scientific literature, it is close to this meaning. It is the time required for the obsolescence of half of the currently published literature (Burton and Kebler, 1960). Half-life implies obsolescence of knowledge, which has significance in scientometric studies.

Several variables are worthy of descriptive and inferential statistical analysis in scientometrics. The variables may refer to the production of scientific knowledge in an institution, a country, a region, such as a continent, or in the world as a whole. These can be further split into subject areas or disciplines.

Varied uses

The major domains of scientometric applications, as confirmed by Gingras (2016), are research evaluation, history of science, sociology of science, science policy, library and information science and economics of science. While scientometrics is functional for the general evaluation of scientific research, it has great potential and possibilities for deeper levels of analysis. Deeper exploration of data can take several layers of analysis using the variables drawn from databases, journals or other documents. The outcomes of such analyses have immediate utilitarian values for an institution or a country in the design of both their short- and long-term research policies. Often, countries rely on the indicators drawn from scientometric data to formulate their science, technology and innovation policies, and to review the production trends in specific fields (Sooryamoorthy, 2020). They are also meaningful for determining the standing of scholars in their respective scientific fields, for purposes of hiring and career advancement.

Researchers who employ scientometrics have found it suitable for a range of other purposes in the study of disciplines. In the early stages of its development, as van Raan (1997) summarises, there were four core interest areas for scientometrics:

i. The development of methods and techniques for the design and application of indicators on important aspects of science and technology. The indicators are meant to measure the performance and impact, international collaboration and specialisation.
ii. The development of information systems on science and technology.
iii. The interaction between science and technology.
iv. The study of the cognitive and socio-organisational structures of scientific fields and developmental processes. This is about scientometrics applied to the sociology of science, citation analysis and mapping.

In the development of scientometrics as an applied method, van Raan (1997) was aware of its future prospects, in particular the work being done by scientometricians in the fields of the HSS and applied sciences. Those discipline-specific studies, as he acknowledged, revealed communication and collaboration features at national, regional and international levels. Scientometric methods were adopted to identify and analyse the role of collaboration networks and the position of the centres of excellence within and outside these networks (van Raan, 1997).

According to Ivancheva (2008), scientometrics relates to the subjects of science by itself in the epistemological sense, scientific knowledge production, and the macro environment of scientific research. In the epistemological sense of science, scientometrics can deal with the general system of development, disciplinary structures and their interrelations along with different types of mathematical model. In the process of involving scientific knowledge production, scientometrics can explain the quantitative characteristics of research potential, communications in science, productivity, evaluation of scientists, institutions and countries, collaboration, communities and networks, coauthorships and funding. For the purpose of the macro environment of scientific research, scientometrics can contribute to science policy, innovation and globalisation as well (Ivancheva, 2008).

Objectives of scientometrics

Theoretically, scientometrics serves two objectives: the relational and the evaluative purposes (Benckendorff, 2009; Benckendorff and Zehrer, 2013). Relational techniques are intended to look at the relationships within research, including the structure of research fields, the emergence of new research themes and methods, and coauthorship patterns (Benckendorff and Zehrer, 2013).

The evaluative purposes encompass the study of the literature used by researchers in any given field. It is about counting references cited by large numbers of researchers in their work (Stevens, 1953). Evaluative uses, as Narin defined them first in 1976, denote the application of techniques such as publication and citation analyses in research assessment. They are applied to the evaluation of scientists, academic departments and scientific publications (Narin, 1976). Evaluative techniques can assess the impact of research, and compare the performance of scientific contributions of individuals and groups (Benckendorff and Zehrer, 2013; van Leeuwen, 2004).

Stevens (1953) put forward a classification concerning descriptive and evaluative focus areas. Descriptive focus areas include the number of publications in any given field, or productivity, which can be used for the purpose of comparing the amount of research produced in countries, periods or in subdivisions of the chosen field. The evaluative one has a focus on research evaluation that can be accomplished from several different perspectives.

In research evaluation, scientometrics was initially employed at the level of countries and institutions, but for some time now it has been applied to evaluate research performance in a variety of scientific fields (Anninos, 2014). Individual scientists and their research were not part of the scheme during the foundational period of scientometrics. Even Garfield thought that the Web of Science (WoS) data, a source of information repository for scientometrics, was not created for measuring the research performance of a scientist. This was, as Gingras (2016) argues, due to the nature of the scientometric data and the fluctuations of the numbers at the micro level of individual analysis. Not much later, the use of the data for the assessment of individual scientists became popular (Gingras, 2016). Universities and research institutions accept the role of research performance and publications of both individuals and teams of scholars in the assessment and allocation of resources. In several Australian universities, for instance, the allocation of research funds is based on a formula that factors in publications (Butler, 2004). In the UK, research ranking, performance measures and funding are well linked (Butler, 2004), showing the influential part of research assessment.

Scientometrics appears to be relevant to the assessment of scientific performance or in the compilation of bibliographies on research domains, but this is not the case. Glänzel (2003) objects to this as a wrong belief. Scientometrics is but a multifaceted endeavour, consisting of several subareas of the structural, dynamic evaluative and predictive scientometrics. Structural scientometrics is about the remapping of the epistemological structure of science, while dynamic scientometrics develops models of scientific growth, obsolescence and citation processes that can be applied in evaluation and prediction (Glänzel, 2003). Note that science studies, as Leydesdorff (2001) explains, is an interdisciplinary field with a variety of epistemological and methodological strands. This makes room for scientometrics to accommodate different paradigmatic approaches as discussed in the previous chapter.

A map of disciplines or research areas, developed through scientometric indicators, is able to identify major areas of disciplines or subjects, their size, similarity and interconnectedness (Boyack et al., 2005). As illustrated in a science map of Japan by Saka and Igami (2014), an overview of the current science and status of the country, categorisation of research areas, and current research activities are gathered.

Scholars have made use of scientometrics to study scientific specialisations pursued in institutions, regions and countries. In one of these kinds of research, Abramo et al. (2014) investigated scientific specialisations in regions and provinces in Italy. By constructing a scientific specialisation index, they found the scientific production of public research organisations in the country. In a related research, the development of science and technology in a newly emerged region of the Union of South American Nations was explored by Greco et al. (2012), who made use of InCites in the WoS data to assess the publication output and citation rates of a few selected countries.

The general concerns of scientometric studies are science and its trajectories of growth and development, which are also part of the evaluative objectives of scientometrics. Alongside these, as its core goal, scientometrics can decompose scientific literature into disciplinary and subdisciplinary structures (Leydesdorff and Rafols, 2009), to investigate the disciplinary trajectories of development. These disciplinary structures are not necessarily confined to science alone.

At this point, it is worth noting some of the key findings of a study on the current trends, objects, concerns and scope of scientometrics. The review of publications in some journals, undertaken by Siluo and Qingli (2017), presents the core content of scientometric studies in the current literature. The five core areas that have been dealt with in the publications are:

i. The extensive application of scientometric methods in biology and other fields.
ii. Research on the quality of research output and research trends of scientometrics as a branch of science.
iii. Research on scientometric methods based on journals and other research outputs.
iv. Development trends of scientific cooperation and networks.
v. Ranking or visualisation of discipline contents through scientometric methods and research on the development trends of disciplines.

These areas, as shown in the ten years of publications between 2007 and 2016, give an idea of the contemporary scope and the foci of scientometrics. As outlined earlier, they fall under the broad areas of applications of scientometrics to science disciplines, namely, research evaluation, collaboration, networks, and mapping of disciplines. These remain the core business of scientometrics.

Some of these central activities deserve a more detailed discussion and elaboration, which is undertaken in the following sections.

Research evaluation and impact

A common use of scientometrics is found in research evaluation, performance and assessment. The rating of research performance on the basis of scientometric indicators is ubiquitous in academic research (Diem and Wolter, 2013). This can be done at the level of an individual, research group/unit, institution, region or country. The tools of scientometrics are extensively applied to the research evaluation of individual scholars, units and institutions as well. In a study of this kind, Moed et al. (1995) presented the essential tools for analysis for the study of fields in the natural and life sciences and the HSS. The tools produced an overall picture of the research performance of individual groups. Statistical analyses can also be performed on the characteristics of the groups. Indicators such as the publication output, trend analysis for short-term impact, impact and cognitive orientation, and collaboration were effectively adopted in this study.

Evaluation models

Developed by some European countries, the research evaluation models have been employed for making informed funding decisions. These models bear an integral component of publications and citations (Kulczycki, 2017). Scientometric indicators give a rationale for the allocation of research funds and quality assurance of research programmes and projects (Diem and Wolter, 2013). Studies of this nature are quite common.

Kulczycki's (2017) critical evaluation of scientific units in Poland produced a relevant model. Equivalent to the one being used in several European countries, the model evaluates scientific units on the basis of four criteria: scientific and creative achievements; scientific potential; material effects of the scientific activity; and other effects of the scientific activity. A few parameters are constructed for each of these criteria which are used differently to suit the nature of the science. Four groups of sciences, namely the social sciences and humanities (SSH), science and engineering (SE), life sciences (LS) and art sciences and artistic production (ASP) are listed.

The parameters for each of the groups comprise of:

1. Scientific and creative achievements:
   SSH: Journal articles, monographs and patents
   SE: Journal articles, monographs and patents
   LS: Journal articles, monographs and patents
   ASP: Journal articles, monographs, patents and artistic production.
2. Scientific potential:
   SSH: Qualifications of academic degrees, promotions and employees' positions in scientific organisations
   SE: Qualifications of academic degrees, promotions, employees' positions in scientific organisations, projects, accredited laboratories and status of the national research institute
   LS: Qualifications of academic degrees, promotions, employees' positions in scientific organisations, accredited laboratories and status of the national research institute
   ASP: Qualifications of academic degrees, promotions and employees' positions in scientific organisations.

3. Material aspects of the scientific activity:
   SSH: Expert opinions and projects
   SE: Expert opinions, salaries from external funds, equipment and software expenses, and commercialisation of technology
   LS: Salaries from external funds, equipment and software expenses, and commercialisation of technology
   ASP: Expert opinions, artistic activities, projects and commercialisation of technology.
4. Other effects of the scientific activity:
   SSH: Prizes, conferences, popularisation of science and other
   SE: Prizes, conferences, popularisation of science and other
   LS: Prizes, conferences, popularisation of science and other
   ASP: Prizes, conferences, popularisation of science and other.

Many of these criteria and their constituent parameters can be obtained from scientometric sources. Articles in scientific journals are then assessed according to the list of journals, prepared annually for the purpose of evaluation. The list includes the journals indexed in the *Journal Citation Reports* (*JCR*), the European Reference Index for the Humanities and the Polish or foreign journals without the impact factor. The points for articles in the list are based on the impact factor of the journal, the number of authors from foreign institutions and the number of articles per year.

Evaluation – steps

The methodological steps to be taken in research evaluation studies are of great importance. Anninos (2014) describes these steps. The research evaluation process, according to Anninos, begins with the determination of the academic level. This is followed by the determination of the period of analysis, databases for the extraction of data and the publication types for analysis. If the evaluation is about a research department (faculty) or a research unit, then the publications of the department and a proper counting mechanism should be identified. Citation analysis is done, and bibliometric indicators are calculated in the final stage. Anninos (2014) realises that certain indicators are more useful than others in research evaluation. They include the number of publications; the number of citations; the average number of citations per publication; the percentage of publications without citations; and the average citation number of all papers published by the department/unit.

Impact assessment – indexes

Impact assessment is as important as research evaluation. On a number of levels, scientometric data can be used for the measurement of the impact of research. Roemer and Borchardt (2015) discuss four levels of metrics to measure the impact of scientific contributions. The individual contributions constitute the first level by supplying the building blocks. These contributions are found in scholarly journals, books, book chapters, blogs, posts, conference proceedings, posters, presentations, datasets, inventions, patents and other creative works. The second level pertains to the venue of production, which could be an academic journal, a book publisher, editor, conference or performance setting. The metrics for the third level refer to individual authors and are measured by the average number of citations, views and the h-index. This may be best suitable for researchers who already have a record of trackable research outputs. In level four, the metrics are meant for groups and institutions, necessitated by the need to measure the impact of groups of researchers, departments and institutions, and are used for funding purposes. The data is aggregated at the macro level.

The invention of the h-index in 2005 by Jorge Eduardo Hirsch (1953–), an Argentine-American physicist, was a major breakthrough in the study of the impact of the work of scholars. Well received in scientometrics (Egghe, 2006), but not without controversy, the index measures the publication productivity of individual scholars. It is defined as "the number of papers with citation number $\geq h$ to characterise the scientific output of a researcher" (Hirsch, 2005: 16569). If a scientist has an index h, and if h of their number of papers have at least h citations, then each of the other papers has $\leq h$ citations (Hirsch, 2005).

In other words, the h-index is calculated by using the number of papers a scholar has already published. If a scholar in the social sciences has published 60 papers during the course of their career, and 15 of these papers have been cited 15 or more times, the h-index of the scholar is 15. The h-index of an author is likely to increase, depending on the number of papers published and the number of citations the papers earned. The greater the number of papers, the more are the chances of citations and increase in the h-index.

The h-index is capable of measuring the broad impact of the work of an individual by eliminating some of the disadvantages of counting the number of papers, citations, citations per paper and the number of significant papers (Hirsch, 2005). It gives an estimate of the importance, significance and broad impact of the cumulative research contributions of a scholar. Relying on the publications in their respective indexes, citation indexes of WoS, namely WoS Core Collection (WoSCC), Scopus and Google Scholar (GS), calculate the h-index of scholars. Although the index is used as a scale to measure the quality of publications, it is neither a measure of quantity nor quality/impact, but a composite of them (Gingras, 2016).

The h-index of Robert Merton, a renowned sociologist who made remarkable contributions to the study of science, is prepared from three sources of data – WoS, Scopus and GS.

Figure 2.1 is prepared from data obtained from WoS. The databases of WoS, namely WoSCC, the KCI-Korean Journal Database, MEDLINE, the Russian Index of Science Citation and the SciELO Citation Index, were used.[2] Depending on the databases, the number of publications and citations vary. Publications in the selected databases cover the period from 1945 to 2019. The diagram shows a total of 80 publications (Merton has authored more than 80, but not all are indexed in the WoSCC database), 6,447 citations and an h-index of 21.

Web of Science

Search   Search Results   Tools ▾   Searches and alerts ▾   Search History   Marked List

Citation report for 80 results from All Databases between   1945 ▾   and   2019 ▾   Go

You searched for: AU=Merton, RK ...More

This report reflects citations to source items indexed within All Databases

Export Data   Save to Excel File   ▾

Total Publications	h-index	Sum of Times Cited	Citing articles
80 Analyze	21	6,447	5,730 Analyze

	Average citations per item	Without self citations	Without self citations
1958 ... 2017	80.59	6,409	5,709 Analyze

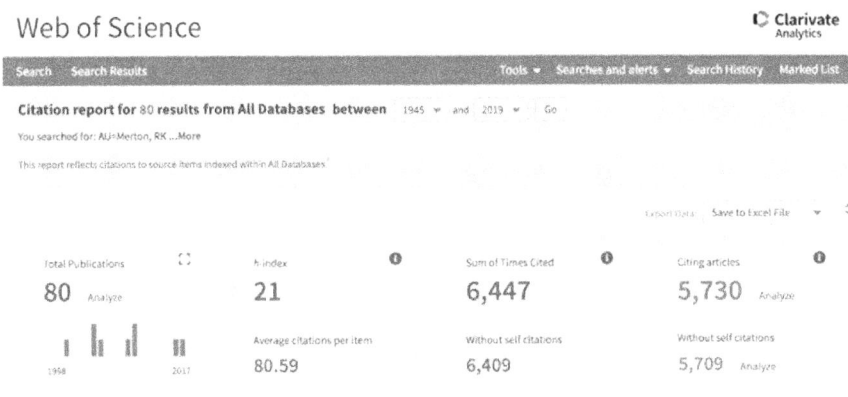

FIGURE 2.1 Citation report of Robert Merton in Web of Science

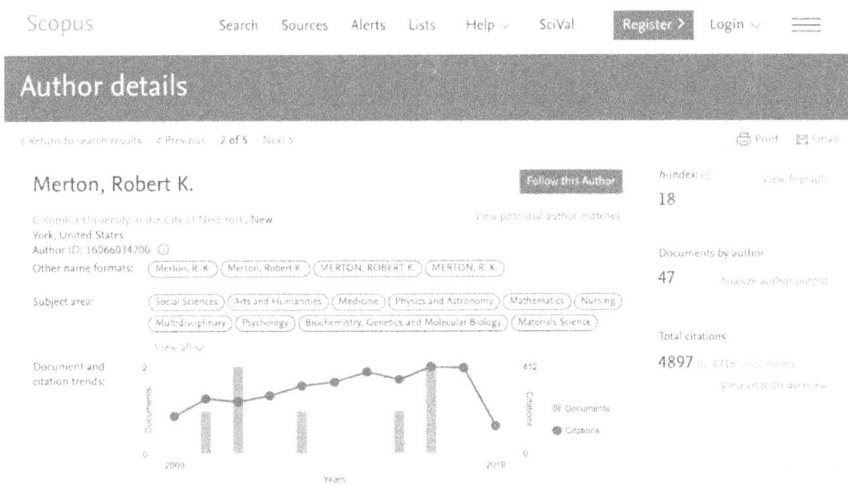

Scopus

Search   Sources   Alerts   Lists   Help ▾   SciVal   Register >   Login ▾   ☰

Author details

‹ Return to search results   ‹ Previous   2 of 5   Next ›   🖨 Print   ✉ Email

Merton, Robert K.

Columbia University in the City of New York, New
York, United States
Author ID: 1606604200 ⓘ

Other name formats: (Merton, R. K.) (Merton, Robert K.) (MERTON, ROBERT K.) (MERTON, R. K.)

Subject area: (Social Sciences) (Arts and Humanities) (Medicine) (Physics and Astronomy) (Mathematics) (Nursing)
(Multidisciplinary) (Psychology) (Biochemistry, Genetics and Molecular Biology) (Materials Science)
View all ▾

Document and
citation trends:

h-index ⓘ
18

Documents by author
47   Analyze author output

Total citations
4897   View citation overview

FIGURE 2.2 Author details of Robert Merton in Scopus

The Scopus database, as shown in the snapshot in Figure 2.2, was searched for the author details of Robert Merton. The page presents subject areas that the publications of Merton fall under, graphical representation of his publications and citations, the *h*-index and the total count of citations. From this page, further navigation is possible to create the *h*-index graph, and to analyse the output of the author. The diagram is based on 47 publications indexed in Scopus, and has 4,897 citations with an *h*-index of 18.

Figure 2.3 shows the data on the profile of Robert Merton, as captured by GS. It presents all the contributions of the author that GS could collect from the internet. On the GS page, the citations of all his publications so far and for the five years since 2014 are graphically demonstrated. Both the count of citations and the *h*-index

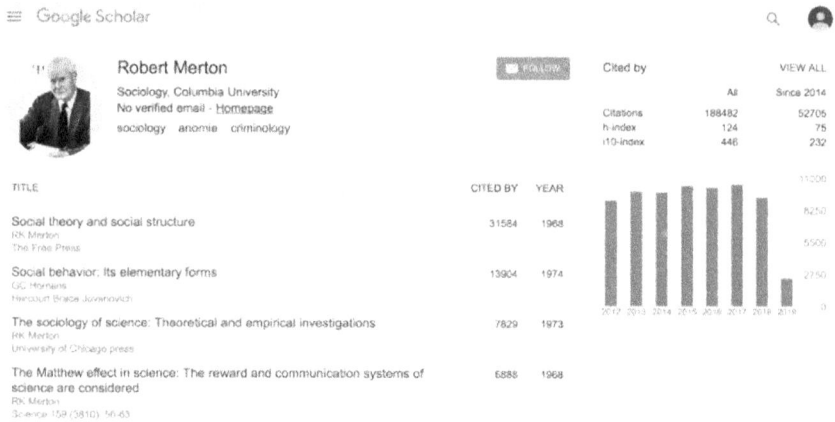

FIGURE 2.3 Author profile of Robert Merton in Google Scholar

are way higher in GS than in WoSCC or Scopus, as it tracks publications that are not covered by WoSCC and Scopus. GS has found a total of 188,482 citations for Merton with an *h*-index of 124. On the GS page, Merton's highest cited paper is shown first. Papers can be sorted according to the year of publication. By moving the cursor on the bar chart, the number of citations for the given years can be revealed.

The differences between WoSCC, Scopus and GS in the count of citations and the *h*-index are due to the coverage of documents tracked and indexed by the respective databases.

Following the acceptance received by the *h*-index, more such indexes were soon to be developed. In 2006, Leo Camiel Jeanettee Egghe, a Belgian mathematician-librarian, proposed an improved measure of the *h*-index. Called the *g*-index, it was created to measure the global citation performance of a set of articles; by definition, "a set of papers has a *g*-index if *g* is the highest rank such that the top *g* papers have, together, at least g^2 citations. This also means that the top $(g + 1)$ papers have less than $(g + 1)^2$ papers" (Egghe, 2006:132). Since then, there have been no shortage of such indexes.

The *h*-index is criticised for many of its drawbacks. Waltman and Eck (2012) argue that in measuring the overall scientific impact of scientists or other units of analysis, the *h*-index behaves in a counterintuitive way. The mechanism used by the *h*-index to aggregate publication and citation statistics into a single number, in certain cases, leads to inconsistencies. Waltman and Eck believe that the *h*-index cannot be used as an appropriate indicator of a scientist's overall scientific impact. The inconsistency of the index has been an issue and scholars have tried to test it (Liu, 2013, for example).

The journal impact factor

Jointly developed by Eugene Garfield and Irving Sher, the journal impact factor (JIF) takes into account relatively large populations of articles and citations (Garfield, 2006). It is a standard tool for measuring the influence and impact a journal has on the scientific community. The JIF is based on two elements. The first is the numerator, which is the number of citations in the current year compared to items published in the previous two years. The second element is the denominator, which is the number of substantive articles and reviews published in the same two years (Garfield, 2006). Citations of a document are measured with a delay of around two years. The JIF is calculated by counting all citations in any given year of all of the papers published in the journal that year, which is then divided by the number of papers in that journal in the previous two years. In other words, the JIF of a journal in 2018 is the number of citations received by all articles in the journal in question in 2018, divided by the total number of articles in the journal between 2016 and 2017.

Garfield thought that if the impact factor could be based on the previous year alone, it would give greater weight to rapidly growing subjects. The two-year period, and not a five- or ten-year period, was chosen, as Garfield at that time did not have more years of the journals he was examining (Ball, 2018). Now, the *JCR* are also available for a five-year period.

The JIF is a measure of quality of a journal, but should not be mistaken as an index of the quality of the articles published in the journal. A higher number of citations do not guarantee the quality of an article. It is a tool for determining the quality of a journal. Journals use this to their marketing advantage as a promotional tool to convince authors to submit their best papers and to sell such journals better than others (Gingras, 2016).

Another core function of scientometrics that is pertinent to the HSS is the mapping of disciplines and subjects.

Mapping of disciplines, subjects and fields

Mapping of scientific fields has been a persistent focus of scientometrics. Similar to the places seen on a geographical map, a map of a discipline, subjects or fields of interest shows the themes and topics in the field. In the map of a discipline or subject, the themes and topics are cognitively related (Noyons, 2004). Maps present structures of scientific literature and underlying specialities (Chen et al., 2002). Mapping is possible in any discipline, subject or field, regardless of its home in science or in the HSS.

The basic unit of analysis in a map is a domain of scientific knowledge in the form of intellectual contributions of the members of a scientific community (Chen, 2017). Mapping requires an assembly of components. These components, as Chen (2017) lists, are the body of scientific literature, scientometrics and visual analytical tools (co-citation analysis, co-word analysis, graphs and network visualisation), metrics and indicators (citation counts, *h*-index and altmetrics) that portray patterns, trends and theories (paradigmatic views, scientific advances driven by competitions and evolutionary stages of disciplines).

Maps are able to answer questions about the subject, discipline or area of inquiry (Noyons, 2004). They show what the domain looks like, who the main actors in the domain are and their particular expertise in the field, how the expertise relates to that of others, the main developments in the area during a particular window period, which actors contribute to these developments, and who may be responsible for a particular change in the area (Noyons, 2004).

While elaborating on mapping research specialities, Morris and Martens (2008) provide a basic model for mapping. They consider a bibliographical approach as one of the four methods (the others being sociological, communicative and cognitive) for mapping research specialities. In this approach, data is collected from written communications in the speciality, which is sourced from web pages of the researchers and institutions, and formal speciality literature. For them, a basic mapping model comprises of three parts: the network of researchers, a system of basic knowledge and a formal literature. These three parts are further broken down into their components and inputs. For instance, researchers need funding input and institutional support for research in a speciality. Similarly, researchers work individually or in teams, which has a communicative component. In the same way, the system of basic knowledge of a speciality requires theories, data, techniques and validation.

Mapping is a rigorous scientific exercise involving appropriate scientometric tools. Some software tools are available for this purpose: VOSViewer and CiteSpace (both freeware) and HistCite are among them. VOSViewer can construct and visualise scientific networks of journals, researchers and individual publications. CiteSpace is able to analyse, detect and visualise trends in scientific literature. It is designed to answer questions relating to major research areas, connections between major areas, the most active research areas in any given field or discipline, and the transition in the history and development of a research field. HistCite is a similar tool, developed by Eugene Garfield, functional for the analysis and visualisation of citation linkages between scientific papers.

In Figure 2.4, the data collected from WoSCC is plotted on a graph. The data represents the count of publications in the discipline of sociology produced in the world during the 19-year period of 2000–2018. The graph indicates publications in

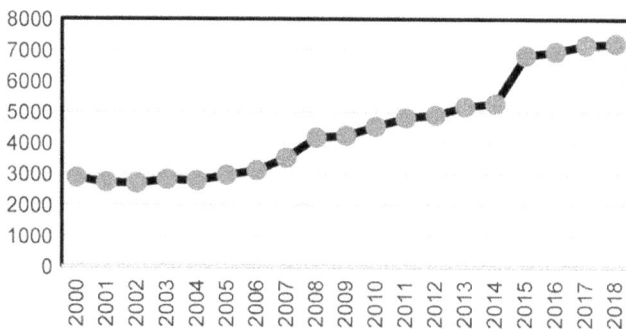

FIGURE 2.4 World production of papers in sociology, 2000–2018
Note: Data is sourced from WoSCC.

No.

32925

1589

FIGURE 2.5 World production of papers in sociology, major countries, 2000–2018
Note: Data is sourced from WoSCC.

sociology in all languages in each of the selected years, which grew from 2000, had a sudden surge in 2015, then flattened out in 2017 and 2018. The same data can be regrouped according to countries, regions or institutions to make further levels of statistical and graphical inferences.

The second image (Figure 2.5) presents the publications in sociology according to the origin of countries on a world map. The intensity of gradience reflects the count of publications, with darker shades meaning a high number of publications and lighter shades indicating a low number of publications. The data illustrates that the USA produced the highest number of publications in sociology during the reference period. The USA is followed by the UK, Canada, Germany, Australia and Russia. These two diagrams show how the data can be plotted graphically.

Co-word analysis

Co-word analysis is a technique for mapping. The structure and dynamics of scientific research can be mapped using this method. Developed at the *École des Mines* and the CNRS (French National Centre for Scientific Research) in Paris in the 1980s, this tool is for finding research fronts by looking at the key terms in the title or abstract in a cluster of articles. Co-word analysis assumes that articles using the same term are related on the cognitive level. The network of co-occurrences between different words, collected from a set of publications, allows a quantitative study of the structure of the publication contents (Tijssen and van Raan, 1989). The method permits the researcher to collect the static and the dynamic aspects of the discipline in which scientists relate and place their work (Tijssen and van Raan, 1989).

Bhattacharya and Basu (1998) show how co-word analysis can be employed to figure out the structure of a speciality. They applied the method to study condensed matter physics (CMP) for two years out of a five-year span between 1990 and 1995. The analysis was performed on the basis of the words extracted from the titles of the selected concurrent set of journals in the subject area. All publications in selected journals were then downloaded, meaningful words from the titles were filtered and a database was created. Words were then ranked according to the frequency of their occurrences. From this list, 50 words were selected on the basis of their maximum occurrence for each year (1990 and 1995) to create co-word pairs. These 50 words represented the most important concepts, which informed research in CMP. The words that are highly ranked show the linkages among the important concepts in CMP. In order to understand the nature of the relationship between the two fields of CMP that emerged in the data, network analysis was applied. Each word is taken as a node, connected by a relation of co-occurrence to another node, which forms the ties/links. Network analysis was then performed on the co-occurrence matrix to map the relationship. The nodes were combined to form blocks, and the relationships between the blocks were then explored. Blocks were classified under plausible research areas in CMP, which were then interpreted as the frontier research areas.

Co-network analysis

First introduced by Jiménez-Contreras et al. (2005) in their study of Spanish psychology journals, co-network analysis uses actors as vectors with their networks already constructed by co-word analysis. The analysis can produce graphic representations in which similar actors of thematic orientation appear in strategic positions.

Co-citation analysis

Corresponding to co-word analysis is the analysis of co-citations. Developed by Henry Small and his colleagues at the then Institute for Scientific Information, co-citation analysis discerns the intellectual structure of science and the connection of speciality areas (Bayer et al., 1990). Small (1973) defined it as the frequency with which two documents are cited together. He realised that co-citation frequency of two papers can be determined by comparing the list of citing documents. The number of identical citing items, according to him, define the strength of co-citation between the two cited papers. An identical citing item is a new document which has cited both earlier papers. This is a relationship established by the citing authors. From the networks of co-cited papers, specific scientific specialities can be traced. Small found in co-citation a new way to study the speciality structure of science. In other words, when a pair of documents is cited together, it is highly likely that the content of the documents is related.

Co-citation analysis has been popular since the early 1980s and is valuable in mapping scientific research and disciplines. It can identify the high-density research fronts in a citation network by clustering highly co-cited documents (Braam et al., 1988, 1991). Co-citation maps reveal clusters of related scientific work (van Raan, 2004). Braam et al. (1991) illustrate the following steps involved in co-citation analysis.

Firstly, documents that are cited more than a specified number of times from the reference list of a set of publications published in any chosen period are selected. From these cited documents, pairs are identified that occur more frequently together in the reference list of publications. These pairs must measure up to some specified co-citation strength threshold. Secondly, a single linkage clustering is constructed. This clustering aggregates groups of cited documents by sequentially linking together all selected pairs of cited documents that have at least one document in common in the publications. Finally, for each cluster, all publications are identified that reference one or more of the clustered cited documents. These identified clusters can be treated as research specialities.

Resonance analysis

Scientometric data can be subjected to resonance analysis. Resonance analysis is not just counting the publications – it is about the perception of the scientific community of the publications (Ball, 2018). The perceptions of the scientific community are ascertained on the basis of the gradual addition to the existing knowledge, as new aspects are added to a question as existing ones are corrected (Ball, 2018). An important indicator of the resonance analysis is the citation frequency, which is the number of citations of a particular publication in a specific period of time. The study of citing publications, particularly the publication that cited the initial paper, can generate a citation analysis on a meta level, revealing which scientists cited the initial publication (Ball, 2018).

Some other forms

Bibliographic coupling is used to identify the related documents in publications. It helps in establishing a similarity relationship between documents. This method links two papers that cite the same paper, and if papers A and B cite paper C, they are related. The more papers A and B cite, the stronger the relationship (Das, 2015). Tijssen et al. (1988) introduced the quasi-correspondence analysis with the objective of mapping the interrelations between journals. The method is based on spatial configurations, resulting from the analysis of citation structure and the distributions of the frequencies of different field-classification codes assigned to documents.

Indexes for mapping

The activity index is a method of studying a research field. It measures the relative strengths of a country's work in any given research field. The index gives a ratio between the share of a country's output in the total world output in a research field, and the share of a country in the world publication output for all science fields (De Bellis, 2009). Attractivity index in terms of citations is another relevant index. It shows the ratio between the share of citations attracted by the publications of a country, and the country's share in citations attracted by publications in all scientific fields combined (De Bellis, 2009). Both are useful in mapping and impact studies.

Immediacy index is a related measure to assess the speed with which articles in a journal are cited. The index is calculated by dividing the number of citations of articles in a journal within a year by the number of articles published in the same year. For example, the immediacy index of a journal that published 50 articles in 2018 and received 500 citations for all the 50 articles is 10. The immediacy effect distinguishes between a literature that is scientific and one that is not. Developed by Price (1965), the Price index is meant to calculate the percentage of references in articles to sources that are not five years older than the citing article. Based on this index, both the "research front" of the discipline which is distinguishable and the citation structure are measured. According to this index, if a subject has approximately 42 per cent of its references that fall within the last five years of the date of the publication of the cited article, it is a hard science. If the percentage is between 42 and 21, it is a soft science, and less than 21 per cent is a non-science.

Specialisation index, or the revealed scientific advantage, is a practical index which gauges the strength of a discipline in a country, region or institution by examining the research output. This is also a measure to find out the importance of a field in relation to other fields. As explained by Archambault and Gagné (2004), the index is represented in two ways. Initially, an aggregate is said to be specialised when it produces more in a specific discipline than in all other disciplines. Secondly, an aggregate is considered as specialised when its percentage of output in any given discipline is higher than the other aggregates. Archambault and Gagné (2004: 32–33) proposed the following formula to calculate the index.

$$IS_{x/y} = (x_a / x_t) / (y_a / y_t) = P_{xa} / P_{ya} \text{ OR}$$
$$IS_{x/y} = (x_a / y_a) / (x_t / y_t) = P_{ax} / P_{ay}$$

Where:
x_a is the number of articles published by group x in discipline a,
y_a is the number of articles published by group y in discipline a,
x_t is the total number of articles published by group x,
y_t is the total number of articles published by reference group y,
P_{xa} is the percentage of articles of group x belonging to discipline a,
P_{ya} is the percentage of articles of group y belonging to discipline a,
P_{ax} is the percentage of articles in discipline a produced by group x, and
P_{ay} is the percentage of articles in discipline a produced by group y.

Group x is always a subset of group y. An index higher than 1 indicates that x is specialised in relation to y, and an index lower than 1 indicates that group x is not specialised in discipline a.

In a similar scenario, but regarding the strengths of a research group, Godin's (2002) index, which was experimented on the social sciences in Canada, takes into account the proportion of a research group, papers in the field and the proportion of papers by all Canadian researchers. The index allows for the measurement of a group of researchers (laboratory, department, institution, region or country) that is more or less specialised or active in a particular field (Godin, 2002). This is in relation to another group of researchers. Godin's formula shows that the index of a specialisation of a group X of Canadian researchers, as compared to the totality of Canadian researchers, is equal to the proportion (%) of researcher group X's papers in field Z *divided by* the proportion (%) of papers by all Canadian researchers in field Z. If the index of specialisation is greater than 1, it means that group X is specialised in the field Z as compared to the reference population, which is Canada.

Studying emerging fields, specialities, concepts and interdisciplinarity

For those who are curious to know about the emerging fields in their own or other disciplines, scientometrics is a reliable method. Several studies have been reported in the literature on this approach to the study of disciplines (Archambault and Gagné, 2004; Morris, 2005). Archambault and Gagné (2004) describe two methods of identifying the emerging fields in a discipline by firstly delineating a speciality and then characterising its development. Method one is to identify the seminal, core articles and then research fronts by way of bibliographic coupling, co-word analysis and co-citation analysis (any one of these or a combination of all three). Based on the growth in the number of publications, it can be estimated whether a field is either emerging or has been stabilised or is a passing fad. Another method is to identify a set of recent articles on a new subject by one or more experts. Using these articles, monitoring is done to detect all exemplars in the field so as to determine whether documents cited in them belong to the research front. The study of academic and content-related trends like this can make predictions on academic and socially relevant topics for strategic decisions (Ball, 2018).

The method is also adopted to analyse the speciality of subjects in disciplines. Rons (2018) had a novel approach to approximate speciality by combining source, title, authors and references from publications. This assists in observing regularities in publication data and applying them to the publications of individual scholars.

The creative application of scientometrics to study disciplines has been found in some recent studies. McKeown, with a team of 20 scholars and specialists, conducted a study of the impact of research concepts that will be promising in future (McKeown et al., 2016). This was done through the analysis of full-text publications instead of the usually adopted metadata of publications to identify the concepts, relations, citation sentiment and the rhetorical function of sentences. In addition to these, the measures obtained from the citation and collaboration networks of authors are supplemented. The evolution of the features of the concepts over time is tracked by this method.

The dynamic relations between disciplines, and their relative openness (for interdisciplinarity) (Gingras, 2016), can be explained with the help of scientometric information and methods. In the recent past, several approaches to the study of interdisciplinary research using scientometrics have emerged and been tested. Such interdisciplinary research can be conducted by looking at the collaboration of authors from different disciplines, co-occurrence of classification codes in publications, interdisciplinary nature of publication journals, and cross-disciplinary references and citations (Bordons et al., 2004). These can be achieved by scientometric methods. Key scientometric indicators useful for the measurement of inter- and multidisciplinarity have been developed and applied (for instance, Morillo et al., 2001; Porter and Chubin, 1985; Sanz-Menéndez et al., 2001; Tomov and Mutafov, 1996).

As part of mapping a discipline or a subject field, the contents of publications are analysed in depth. Murray et al. (2019) developed a method of investigating the disputes and disagreements in publications. Although their focus is on science publications, the method is applicable to the HSS due to the presence of dispute and disagreements in the HSS literature. Signal phrases of disagreement such as contradict, conflict, studies and results are assessed to find out the prevalence of disagreement across publications and disciplines. The method can reveal the position of disagreement in publications (introduction/discussion) as well.

When scientometric methods function as apt tools to study both overt and covert aspects of disciplines and subjects, their applications are being creatively experimented with in new areas. Such efforts are indicative of the growing potential of the methods in science and the HSS. The production of knowledge occurs in a range of locations. Individuals, research teams, institutions, countries, regions and continents are the points of origin of knowledge. It is not surprising if someone is curious about how knowledge travels from point to point. Can a scientometric study reveal the flow of knowledge from place to place? Yes, if one follows the new approach developed by Abramo and D'Angelo (2019). In accepting that all new knowledge produced and measured by publications is indexed in citation databases, they correctly assume citations are proxies of scholarly impact. Based on these premises, Abramo and D'Angelo measured the spillover of knowledge across regions. Relying on the WoSCC data, the study measures the outflows of knowledge produced in one region to other regions in a country (Italy), and the inflows of knowledge produced by other areas in a region. This model explains the scope and applications for the study in larger regions, or from country to country, to understand the dynamics of knowledge production, knowledge flow and knowledge transfer.

Coauthorship, collaboration and networking

Studies of coauthorships, collaboration and networking are prolific in scientometric literature. Given the volume of studies in this domain, they are perhaps some of the major applications and uses of scientometrics. The extent of collaboration in science, as viewed by Subramanyan (1983), cannot be easily determined by traditional methods, but scientometric methods are a convenient tool. Scientometric studies are beneficial for gathering patterns in collaboration and forms of research networks, specific to disciplines, institutions and countries.

Uses

Collaboration and network studies rely on the types of coauthorship obtained from publication records. Coauthorship is a proxy to scientific collaboration and networks. Researchers work together on research projects and the outcome of the project often appears as a publication under joint authorship. From such joint authorship, dimensions of research collaboration can be inferred and analysed. Although coauthorships and research partnerships are more prevalent in science disciplines, they are not rare in the HSS. Because of the benefits accrued by authors and institutions, coauthorship, and by implication collaboration and networks, has importance and value in science enterprise. Researchers have found the analysis of collaboration in the HSS to be worthwhile.

Forms of collaboration

Coauthorship gives rise to several forms of collaboration. Collaboration can be within or between a department, institution or countries. It may be sometimes specific to subjects, research areas and disciplines. Collaboration is defined as internal (within department, discipline, institution or country) or external (between departments, disciplines, institutions or countries) or both. They can also be categorised broadly into domestic and international collaboration. Domestic collaboration is when authors come from the same country, while international collaboration, as the name indicates, occurs when authors join from two or more different countries.

Sources of information

For the purpose of collaboration and network studies, information is drawn primarily from authors' affiliation addresses, which are often found in the metadata of publication records. The raw information from authors' affiliation is collected manually from citation indexes or other publication sources before it is captured in a software program for detailed statistical analysis. In order to know the intensity of collaboration, one needs to know how many authors were involved in the production of a publication, which should be counted manually. From this information it is easy to calculate other useful measures. Fractional count of authors, for example, is calculated by dividing the count of publications by the number of authors. Such measures contribute to a more detailed analysis of collaboration and networks.

The count of countries is of great advantage to collaboration and network studies. This information, like that of the count of authors, is sourced from the affiliation addresses of authors. Here again, this is to be counted manually before it can be processed for further analysis. In other words, a researcher who is keen on understanding collaboration or networks of authors should be looking for the count of authors in publications, their disciplinary or departmental background, institutional affiliation and country of the institution. From these basic variables, more complex analyses can be performed by transforming the basic variables.

Scope of analysis

Scientometric analyses may be focused on a number of levels – individual, disciplinary, institutional, regional, international – or at all levels. Studies of collaboration have implications for subject areas, disciplines, institutions and countries, and can reveal many interesting dimensions of coauthorship, collaboration and networks.

In some countries, one may not find much domestic collaboration (between authors affiliated to institutions in the same country), but more inclination towards international collaboration (collaboration between countries). The extent of international collaboration is evident in the number of participating countries in a publication.

Measuring collaboration

Indexes have been developed to measure the extent and strength of collaboration (Sooryamoorthy, 2015; Wang et al., 2017) to find out the relationship between collaboration and other relevant variables. Such indexes serve well when advanced statistical procedures are to be employed. The degree of collaboration, a basic index, is measured with scientometric data. The degree of collaboration is referred to as a ratio of the number of collaborative research papers to the total number of research papers published in the discipline, during a chosen period of time (Subramanyan, 1983).[3]

Relationships

Collaboration can be connected to disciplines, subjects and/or research areas. Some disciplines or subject areas might be more collaborative than others. Interdisciplinary collaboration can be assessed by the departmental or disciplinary affiliation of authors.

The close relationship between coauthorship and impact of research has been shown in studies (Bornmann, 2017; Gazni et al., 2016; Guan et al., 2017; Katz and Hicks, 1997; Khor and Yu, 2016; Xia et al., 2014, to cite a few). Compared to sole-authored publications, coauthored publications have a higher likelihood of attracting citations. Increasing collaboration tendencies across institutions and countries, regardless of individuals, institutions, disciplines, regions and countries, make scientometrics a handy tool for investigation. Disciplinary differences, as Puuska et al. (2014) show, influence citations. Choosing a single discipline,

Han et al. (2014) investigated the trends in collaboration and networks at both institutional and country levels. Certain forms of collaboration such as international collaboration have a greater impact than other forms. Similarly, collaboration has a positive correlation with the productivity of scientists (Lee and Bozeman, 2005).

Sample studies

In one of the earlier studies using scientometric data, international collaboration in the European Union (EU) was studied by Glänzel et al. (1999). At the international level, Wang et al. (2017) examined collaboration patterns that existed between China and the member states of the EU. The evolution of research collaboration has been a consistent focus in scientometric studies (Bellotti et al., 2016; Coccia and Bozeman, 2016, for instance). New dimensions of coauthorships are being examined using scientometric data. Stefano and Zaccarin (2016) explored the effects of the collaborative behaviour of researchers on their scientific performance. The social dynamics of coauthorships across disciplines has been investigated in the study that Tsai et al. (2016) conducted. Maisonobe et al. (2016) were attracted by the evolution of world collaboration networks. The dynamics of core-periphery structure in collaboration networks interested Karlovčec et al. (2016). Sangam and Keshava (2005) illustrated Indian collaboration and authorship patterns in some of the social science disciplines.

In network analysis, studies have looked at the structure and influence of collaborative networks, networks across groups and academic fields (Kyvik and Reymert, 2017; Landini et al., 2015; Muniz et al., 2015; Wagner and Leydesdorff, 2005). Muniz et al. (2015), in their analysis of coauthorship, revealed the structure of relational networks of authors in academic research. The structure and dynamics of networks that are specific to international collaborations in northern Africa were of interest to Landini et al. (2015). Using both scientometric and social network methods, an international collaboration of Chinese researchers was viewed from the perspective of authors, countries, discipline and journals (Niu and Qiu, 2014). The evolution of social networks in scientific collaborations has been traced in some chosen disciplines (Barabasi et al., 2002). Networks of scientists have been created on the basis of coauthorship and collaboration (Newman, 2004). Yoshikane and Kageura (2004) attempted a comparative analysis of coauthorship networks in some scientific domains. The literature on collaboration and networks is expansive and is continually growing in the scientometric field.

Big data and scientometrics

In the early years of scientometrics and up until the 1950s, scientometric analyses were restricted to small sets of data, as they were not computerised in the way that they are today. Analyses were performed offline. Indexes were only in print form and were later produced in CD formats. The situation began to change after the

1980s when the data was online, which facilitated the extension of analysis to a larger scale.

Owing to the manual collection and storage of data, earlier studies in the field were of small scale in nature with limited objectives (Sugimoto and Larivière, 2018). With the advancements and developments in the cyberworld, where online resources are more common, accessible and popular, the analysis of huge quantities of data is realisable.

The growing list of data sources contributes to the growth of scientometrics. WoS, Scopus, GS and SCImago are prominent databases that are adapted extensively. The first three are nevertheless the most important multidisciplinary databases providing metadata on scientific documents and citation links between documents (Visser et al., 2019).

Databases store millions of data records, which are growing in capacity year after year, making scientometrics an effective tool for the analysis of big data. Big data, as Jin et al. (2015) characterise, is huge in volume, high in velocity, high in variety, low in veracity and high in value. Indexing databases such as WoSCC and Scopus are in the big metadata domain (Bratt et al., 2017). Big data is not inclusive to the domain of computer science alone, but also comes from disciplines such as biological and medical sciences, management, health care and library and information science (Singh et al., 2015; Xian and Madhavan, 2014).

The hugeness of data is evident from the details of databases. One of the major databases, Clarivate Analytics' WoS, grew from 700,000 documents in 1980 to 2 million documents in 2010 (Cantú-Ortiz, 2018). This database, as of 3 September 2014, held around 50,000 books, 12,000 journals and 160,000 conference proceedings, with 90 million total records and 1 billion total cited references (Cantú-Ortiz, 2018). By 5 February 2019, the WoS platform had grown to a reserve of over 155 million records and over 34,200 journals.[4] Elsevier's Scopus was launched in 2004, and by 2017 it had 67 million items sourced from more than 22,500 serial titles, 96,000 conferences and 136,000 books from over 7,500 publishers (Cantú-Ortiz, 2018). GS, also started in 2004, is a free service providing author citations. Although there are resemblances in the purposes of these three databases, GS cannot be treated like WoSCC or Scopus. The latter have more advanced uses as far as scientometrics is concerned. These data repositories have become de facto measures for assessing scholarly performance (Tang et al., 2013).

The data in repositories, as seen from the above figures, is enormous. In view of the possibilities and potential of the indicators that can be collected from millions of records, scientometrics is an attractive method for students and researchers in big data analysis. The raw variables in these databases can be transformed to create many more meaningful variables. This advantage makes the data suitable for the type of analysis the researcher is seeking, and to gain insights into the dynamics of the production of knowledge and communication. Big data takes scientometrics to a higher level of analysis of finding correlations rather than just causality (Ball, 2018). On the other hand, scientometric research on big data itself is also possible. Singh et al. (2015) made an analysis of research done on big data. Xian and Madhavan (2014) performed big data analysis in engineering education. In big data, the possibilities for scientometrics are bigger. In the next chapter, scientometrics for the HSS are discussed.

Notes

1 Alan Pritchard was critical of the use of the term statistical bibliography, first mentioned by Wyndham Hulme in his lecture at the University of Cambridge in 1922, to mean the illumination of the process of science and technology by counting documents (Hulme, 1923). The term was ignored for several years. Instead, Pritchard coined a new term, bibliometrics, which meant the application of mathematics and statistical methods to books and other media of communication (Pritchard, 1969).
2 Note that these are not all the databases of WoS. Many more exist. Access to these depends on subscription (institution or individual), which is expensive.
3 The formula is C (the degree of collaboration in a discipline) = N_m (the number of multiple-authored papers in a discipline in a year) / N_m + N_c (the number of single-authored papers in a discipline in a year).
4 These figures are as of 14 March 2019 (https://clarivate.libguides.com/woscc/coverage, accessed 19 April 2019).

References

Abramo, G. & D'Angelo, C. A. (2019). The regional balance of knowledge flows. In G. Catalano, C. Daraio, M. Gregori, H. F. Moed & G. Ruocco (Eds), *Proceedings of the 17th Conference of the International Society for Scientometrics and Informetrics* (Vol. 1, pp. 223–234). Rome: Edizioni Efesto.

Abramo, G., D'Angelo, C. A. & Costa, F. D. (2014). A new bibliometric approach to assess the scientific specialization of regions. *Research Evaluation, 23*, 183–194. DOI: 10.1093/reseval/rvu005

Andrés, A. (2009). *Measuring Academic Research: How to Undertake a Bibliometric Study*. Oxford: Chandos Publishing.

Anninos, L. N. (2014). Research performance evaluation: Some critical thoughts on standard bibliometric indicators. *Studies in Higher Education, 39*, 1542–1561. https://doi.org/10.1080/03075079.2013.801429

Archambault, É. & Gagné, É.V. (2004). *The Use of Bibliometrics in the Social Sciences and Humanities*. Montreal: Social Sciences and Humanities Research Council of Canada (SSHRCC). www.science-metrix.com/pdf/SM_2004_008_SSHRC_Bibliometrics_Social_Science.pdf

Ball, R. (2018). *An Introduction to Bibliometrics: New Development and Trends*. Cambridge, MA: Chandos Publishing.

Barabasi, A. L., Jeong, H., Neda, Z., Ravasz, E., Schubert, A. & Vicsek, T. (2002). Evolution of the social network of scientific collaborations. *Physica A: Statistical Mechanics and its Applications, 311*, 590–614. https://doi.org/10.1016/S0378-4371(02)00736-7

Bayer, A. E., Smart, J. C. & McLaughlin, G. W. (1990). Mapping intellectual structure of a scientific subfield through author cocitations. *Journal of the American Society for Information Science, 41*, 444–452. DOI: 10.1002/(SICI)1097–4571(199009)41:6%3C444::AID-ASI12%3E3.0.CO;2-J

Bellotti, E., Kronegger, L. & Guadalupi, L. (2016). The evolution of research collaboration within and across disciplines in Italian Academia. *Scientometrics, 109*, 783–811. https://doi.org/10.1007/s11192-016-2068-1

Benckendorff, P. (2009). Themes and trends in Australian and New Zealand tourism research: A social network analysis of citations in two leading journals (1994–2007). *Journal of Hospitality and Tourism Management, 16*, 1–15. https://doi.org/10.1375/jhtm.16.1.1

Benckendorff, P. & Zehrer, A. (2013). A network of analysis of tourism research. *Annals of Tourism Research, 13*, 121–149. https://doi.org/10.1016/j.annals.2013.04.005

Bhattacharya, S. & Basu, P. K. (1998). Mapping a research area at the micro level using co-word analysis. *Scientometrics, 43*, 359–372. https://doi.org/10.1007/BF02457404

Bordons, M., Morillo, F. & Gómez, I. (2004). Analysis of cross-disciplinary research through bibliometric tools. In H. F. Moed, W. Glänzel & U. Schmoch (Eds), *Quantitative Science and Technology Research: The Use of Publication and Patent Statistics in Studies of S&T Systems* (pp. 437–456). New York: Kluwer Academic Publishers. https://doi.org/10.1007/1-4020-2755-9_2

Bornmann, L. (2017). Is collaboration among scientists related to the citation impact of papers because their quality increases with collaboration? An analysis based on data from F1000Prime and normalized citation scores. *Journal of the American Society for Information Science and Technology, 68*(4), 1036–1047. https://doi.org/10.1002/asi.23728

Boyack, K. W., Klavans, R. & Böner, K. (2005). Mapping the backbone of science. *Scientometrics, 64*, 351–374. https://doi.org/10.1007/s11192-005-0255-6

Braam, B. R., Moed, H. F. & Raan, A. F. J. v. (1988). Mapping of science: Critical elaboration and new approaches, a case study in agricultural biochemistry. In L. Egghe & R. Rousseau (Eds), *Informetrics 87/88: Select Proceedings of the First International Conference on Bibliometrics and Theoretical Aspects of Information Retrieval, Diepenbeek, Belgium, 25–28 August 1987* (pp. 15–28). Amsterdam: Elsevier Science Publishers.

Braam, R. F., Moed, H. F. & Raan, A. F. J. v. (1991). Mapping of science by combined co-citation and word analysis. I. Structural aspects. *Journal of the American Society for Information Science, 42*, 233–251. DOI: 10.1002/(SICI)1097–4571(199105)42:43.0.CO;2-I

Bratt, S., Hemsley, J., Qin, J. & Costa, M. (2017). Big data, big metadata and quantitative study of science: A workflow model for big scientometrics. *Proceedings of the Association for Information Science and Technology, 54*, 36–45. https://doi.org/10.1002/pra2.2017.14505401005

Broadus, R. N. (1987). Early approaches to bibliometrics. *Journal of the American Society for Information Science and Technology, 38*, 127–129. https://doi.org/10.1002/(SICI)1097–4571(198703)38:2<127::AID-ASI6>3.0.CO;2-K

Burton, R. E. & Kebler, R. W. (1960). The "half-life" of some scientific and technical literatures. *Journal of the Association for Information Science and Technology, 11*, 1–87. https://doi.org/10.1002/asi.5090110105

Butler, L. (2004). What happens when funding is linked to publication counts? In H. F. Moed, W. Glänzel & U. Schmoch (Eds), *Quantitative Science and Technology Research: The Use of Publication and Patent Statistics in Studies of S&T Systems* (pp. 389–405). New York: Kluwer Academic Publishers. https://doi.org/10.1007/1-4020-2755-9_18

Cantú-Ortiz, F. J. (2018). Data analytics and scientometrics: The emergence of research analytics. In F. J. Cantú-Ortiz (Ed.), *Research Analytics: Boosting University Productivity and Competitiveness through Scientometrics* (pp. 1–11). Boca Raton, FL: CRC Press.

Chen, C. (2017). Science mapping: A systematic review of the literature. *Journal of Data and Information Sciences, 2*, 1–40. DOI: 10.1515/jdis-2017-0006

Chen, C., McCain, K., White, H. & Lin, X. (2002). Mapping scientometrics (1981–2001). *Proceedings of the American Society for Information Science and Technology, 39*, 25–34. https://doi.org/10.1002/meet.1450390103

Coccia, M. & Bozeman, B. (2016). Allometric models to measure and analyze the evolution of international research collaboration. *Scientometrics, 108*, 1065–1084. https://doi.org/10.1007/s11192-016-2027-x

Das, A. K. (2015). Introduction to research evaluation metrics and related indicators. In S. Mishra & B. K. Sen (Eds), *Open Access for Researchers, Module 4: Research Evaluation Metrics* (pp. 5–18). Paris: UNESCO. https://unesdoc.unesco.org/ark:/48223/pf0000232210

De Bellis, N. (2009). *Bibliometrics and Citation Analysis: From the Science Citation Index to Cybermetrics.* Lanham, MD: The Scarecrow Press, Inc.

Diem, A. & Wolter, S. C. (2013). The use of bibliometrics to measure research performance in education sciences. *Research in Higher Education, 54*, 86–114. https://doi.org/10.1007/s11162-012-9264-5

Egghe, L. (2006). Theory and practise of the *g*-index. *Scientometrics, 69*, 131–152. https://doi.org/10.1007/s11192-006-0144-7

Garfield, E. (2006). The history and meaning of the journal impact factor. *Journal of American Medical Association, 295*, 90–93. DOI: 10.1001/jama.295.1.90

Garfield, E. (2007). The evolution of the Science Citation Index. *International Microbiology, 10*, 65–90. DOI: 10.2436/20.1501.01.10

Gazni, A., Lariviére, V. & Didegah, F. (2016). The effect of collaborators on institutions' scientific impact. *Scientometrics, 109*, 1209–1230. https://doi.org/10.1007/s11192-016-2101-4

Gingras, Y. (2016). *Bibliometrics and Research Evaluation: Uses and Abuses.* Cambridge, MA: MIT Press.

Glänzel, W. (2003). Bibliometrics as a research field: A course on theory and application of bibliometric indicators. Retrieved from http://yunus.hacettepe.edu.tr/~tonta/courses/spring2011/bby704/bibliometrics-as-a-research-field-Bib_Module_KUL.pdf 21 May 2019.

Glänzel, W., Schubert, A. & Czerwon, H. J. (1999). A bibliometric analysis of international scientific cooperation of the European Union (1985–1995). *Scientometrics, 45*, 185–202. https://doi.org/10.1007/BF02458432

Godin, B. (2002). *The Social Sciences in Canada: What Can We Learn from Bibliometrics. Working Paper No. 1.* Quebec: INRS, University of Quebec. www.csiic.ca/PDF/CSIIC.pdf

Greco, A., Bornmann, L. & Marx, W. (2012). Bibliometric analysis of scientific development in countries of the Union of South American Nations (UNASUR). *El profesional de la información, 21*, 607–612. DOI: 10.3145/epi.2012.nov.07

Guan, J., Yan, Y. & Zhang, J. J. (2017). The impact of collaboration and knowledge networks on citations. *Journal of Informetrics, 11*, 407–422. DOI: 10.1016/j.joi.2017.02.007

Han, P., Shi, J., Li, X., Wang, D., Shen, S. & Su, X. (2014). International collaboration in LIS: Global trends and networks at the country and institution level. *Scientometrics, 98*, 53–72. https://doi.org/10.1007/s11192-013-1146-x

Hirsch, J. E. (2005). An index to quantify an individual's scientific research output. *Proceedings of the National Academy of Sciences, 102*, 16569–16572. https://doi.org/10.1073/pnas.0507655102

Hulme, W. (1923). *Statistical Bibliography in Relation to the Growth of Modern Civilization: Two Lectures Delivered in the University of Cambridge in May, 1922.* London: Butler & Tanner Grafton Co. https://doi.org/10.1038/112585a0

Ivancheva, L. (2008). Scientometrics today: A methodological overview. *Collnet Journal of Scientometrics and Information Management Decision, 2*, 47–56. https://doi.org/10.1080/09737766.2008.10700853

Jiménez-Contreras, E., Delgado-López-Cózar, E., Ruiz-Pérez, R., Moneda-Corrochano, M. d. l., Ruiz-Baños, R. & Bailón-Moreno, R. (2005). Co-network analysis. In P. Ingwersen & B. Larsen (Eds), *Proceedings of ISSI 2005 – the 10th International Conference of the International Society for Scientometrics and Informetrics* (Vol. 1, pp. 417–425). Stockholm: Karolinska University Press.

Jin, X., Wah, B. W., Cheng, X. & Wang, Y. (2015). Significance and challenges of big data research. *Big Data Research*, *2*, 59–64. DOI: 10.1016/j.bdr.2015.01.006

Karlovčec, M., Lužar, B. & Mladenić, D. (2016). Core-periphery dynamics in collaboration networks: The case study of Slovenia. *Scientometrics*, *109*, 1561–1578. https://doi.org/10.1007/s11192-016-2154-4

Katz, J. S. & Hicks, D. (1997). How much a collaboration worth? A calibrated bibliometric model. *Scientometrics*, *40*, 541–554. DOI: 10.1007/BF02459299

Khor, K. A. & Yu, L.-G. (2016). Influence of international co-authorship on the research citation impact of young universities. *Scientometrics*, *107*, 1095–1110. https://doi.org/10.1007/s11192-016-1905-6

Kulczycki, E. (2017). Assessing publications through a bibliometric indicator: The case of comprehensive evaluation of scientific units in Poland. *Research Evaluation*, *26*, 41–52. https://doi.org/10.1093/reseval/rvw023

Kyvik, S. & Reymert, I. (2017). Research collaboration in groups and networks: Differences across academic fields. *Scientometrics*, *113*, 951–967. https://doi.org/10.1007/s11192-017-2497-5

Landini, F., Malerba, F. & Mavilia, R. (2015). The structure and dynamics of networks of scientific collaborations in Northern Africa. *Scientometrics*, *105*, 1787–1807. https://doi.org/10.1007/s11192-015-1635-1

Lee, S. & Bozeman, B. (2005). The impact of research collaboration on scientific productivity. *Social Studies of Science*, *35*, 673–702. https://doi.org/10.1177/0306312705052359

Leydesdorff, L. (1987). Various methods for the mapping of science. *Scientometrics*, *11*, 295–324. https://doi.org/10.1007/BF02279351

Leydesdorff, L. (2001). *The Challenge of Scientometrics: The Development, Measurement, and Self-Organization of Scientific Communities* (Second ed.). Irvine, CA: Universal Publishers. DOI: 10.2139/ssrn.3512486

Leydesdorff, L. & Rafols, I. (2009). A global map of science based on the ISI subject categories. *Journal of the American Society for Information Science and Technology*, *60*, 348–362. https://doi.org/10.1002/asi.20967

Liu, Y. (2013). To what extent is the *h*-index inconsistent? Is strict consistency a reasonable requirement for a scientometric indicator? In J. Gorraiz, E. Schiebel, C. Gumpenberger, M. Hörlesberger & H. Moed (Eds), *14th International Society of Scientometrics and Informetrics Conference* (Vol. 2, pp. 1696–1710). Vienna: AIT Austrian Institute of Technology GmbH.

Maisonobe, M., Eckert, D., Grossetti, M., Jégou, L. & Milard, B. (2016). The world network of scientific collaborations between cities: Domestic or international dynamics? *Journal of Informetrics*, *10*, 1025–1036. https://doi.org/10.1016/j.joi.2016.06.002

McKeown, K., III, H. D., Chaturvedi, S., Paparrizos, J., Thadani, K., Barrio, P., … Neelakantan, A. (2016). Predicting the impact of scientific concepts using full-text features. *Journal of the American Society for Information Science and Technology*, *67*, 2684–2696. https://doi.org/10.1002/asi.23612

Moed, H. F., Bruin, R. E. D. & van Leeuwen, T. N. (1995). New bibliometric tools for the assessment of national research performance: Database description, overview of indicators and first applications. *Scientometrics*, *33*, 381–422. https://doi.org/10.1007/BF02017338

Morillo, F., Bordons, M. & Gómez, I. (2001). An approach to interdisciplinarity through bibliometric indicators. *Scientometrics*, *51*, 203–222. https://doi.org/10.1023/A:1010529114941

Morris, S. A. (2005). Manifestation of emerging specialties in journal literature: A growth model of papers, references, exemplars, bibliographic coupling, cocitation, and clustering coefficient distribution. *Journal of the American Society for Information Science and Technology*, *56*, 1250–1273. https://doi.org/10.1002/asi.20208

Morris, S. A. & Martens, B. V. d. V. (2008). Mapping research specialties. *Annual Review of Information Science and Technology*, *42*, 213–295. https://doi.org/10.1002/aris.2008.1440420113

Muniz, N. M., Ariza-Montes, J. A. & Molina, H. (2015). How scientific links combine to thrive academic research in universities: A social network analysis approach on the generation of knowledge. *Asia-Pacific Educational Research*, *24*, 613–623. https://doi.org/10.1007/s40299-014-0207-0

Murray, D., Lamers, W., Boyack, K., Larivière, V., Sugimoto, C. R., Eck, N. J. v. & Waltman, L. (2019). Measuring disagreement in science. In G. Catalano, C. Daraio, M. Gregori, H. F. Moed & G. Ruocco (Eds), *Proceedings of the 17th Conference of the International Society for Scientometrics and Informetrics* (Vol. II, pp. 2370–2376). Rome: Edizioni Efesto.

Narin, F. (1976). *Evaluative Bibliometrics. The Use of Publications and Citation Analysis in the Evaluation of Scientific Activity*. Cherry Hill, NJ: Computer Horizons, Inc.

Newman, M. E. J. (2004). Coauthorship networks and patterns of scientific collaboration. *Proceedings of the National Academy of Sciences of the United States of America*, *101*, 5200–5205. https://doi.org/10.1073/pnas.0307545100

Niu, F. & Qiu, J. (2014). Network structure, distribution and the growth of Chinese international research collaboration. *Scientometrics*, *98*, 1221–1233. https://doi.org/10.1007/s11192-013-1170-x

Noyons, E. C. M. (2004). Science maps within a science policy context: Improving the utility of science and domain maps within a science policy and research management context. In H. F. Moed, W. Glänzel & U. Schmoch (Eds), *Quantitative Science and Technology Research: The Use of Publication and Patent Statistics in Studies of S&T Systems* (pp. 237–255). New York: Kluwer Academic Publishers. https://doi.org/10.1007/1-4020-2755-9_11

Porter, A. L. & Chubin, D. E. (1985). An indicator of cross-disciplinary research. *Scientometrics*, *8*, 161–176. https://doi.org/10.1007/BF02016934

Price, D. J. d. S. (1965). Networks of scientific papers. *Science*, *149*(3683), 510–515.

Pritchard, A. (1969). Statistical bibliography or bibliometrics? *Journal of Documentation*, *25*, 348–349.

Pritchard, A. & Witting, G. R. (1981). *Bibliometrics: A Bibliography of Index*. Watford, UK: ALLM Books.

Puuska, H.-M., Muhonen, R. & Leino, Y. (2014). International and domestic co-publishing and their citation impact in different disciplines. *Scientometrics*, *98*, 823–839. https://doi.org/10.1007/s11192-013-1181-7

Roemer, R. C. & Borchardt, R. (2015). *Meaningful Metrics: A 21st-Century Librarian's Guide to Bibliometrics, Altmetrics, and Research Impact*. Chicago, IL: Association of College and Research Libraries.

Rons, N. (2018). Bibliometric approximation of a scientific specialty by combining key sources, title words, authors and references. *Journal of informetrics*, *12*, 113–132. https://doi.org/10.1016/j.joi.2017.12.003

Saka, A. & Igami, M. (2014). *Science Map 2010 & 2012: Study on Hot Research Area (2005–2010 and 2007–2012) by Bibliometric Method*. Report No. 159. Tokyo: Research Unit for Science and Technology Analysis and Indicators National Institute of Science and Technology Policy (NISTEP).

Sangam, S. L. & Keshava. (2005). Collaboration in social science research in India. In P. Ingwersen & B. Larsen (Eds), *Proceedings of ISSI 2005 – The 10th International Conference of the International Society for Scientometrics and Informetrics* (Vol. 1, pp. 775–778). Stockholm: Karolinska University Press.

Santarem, L. G. d. S. & Oliveira, E. F. T. d. (2009). Co-citation analysis on subject treatment of information: A study based on publications from Brazilian researchers on information science. In B. Larsen & J. Leta (Eds), *Proceedings of ISSI 2009 – The 12th International Conference of the International Society for Scientometrics and Informetrics* (Vol. 2, pp. 990–991). São Paulo, Brazil: BIREME/PAHO/WHO and Federal University of Rio de Janeiro.

Sanz-Menéndez, L., Bordons, M. & Zulueta, M. A. (2001). Interdisciplinarity as a multidimensional concept: Its measure in three different research areas. *Research Evaluation, 10*, 47–58. DOI: 10.3152/147154401781777123

Siluo, Y. & Qingli, Y. (2017). *Are Scientometrics, Informetrics, and Bibliometrics different? Conference Proceedings of the 16th International Conference on Scientometrics & Informetrics* (pp. 1507–1518). Wuhan, China.

Singh, V. K., Banshal, S. K., Singhal, K. & Uddin, A. (2015). Scientometric mapping of research on "Big Data". *Scientometrics, 105*, 727–741. https://doi.org/10.1007/s11192-015-1729-9

Small, H. (1973). Co-citation in the scientific literature: A new measure of the relationship between two documents. *Journal of the American Society for Information Science, 24*, 265–269. DOI: 10.1002/asi.4630240406

Sooryamoorthy, R. (2015). *Transforming Science in South Africa: Development, Collaboration and Productivity*. Hampshire and New York: Palgrave Macmillan.

Sooryamoorthy, R. (2020). *Science, Policy and Development in Africa: Challenges and Prospects*. London: Cambridge University Press.

Stefano, D. D. & Zaccarin, S. (2016). Co-authorship networks and scientific performance: An empirical analysis using the generalized extreme value distribution. *Journal of Applied Statistics, 43*, 262–279. https://doi.org/10.1080/02664763.2015.1017719

Stevens, R. E. (1953). *Characteristics of Subject Literatures*. Chicago, IL: American College and Research Library Monography Series.

Subramanyan, K. (1983). Bibliometric studies of research collaboration: A Review. *Journal of Information Science, 6*, 33–38. https://doi.org/10.1177/016555158300600105

Sugimoto, C. R. & Larivière, V. (2018). *Measuring Research: What Everyone Needs to Know*. New York: Oxford University Press.

Tang, M.-c., Wang, C.-m., Chen, K.-h. & Hsiang, J. (2013). Exploring alternative cyberbibliometrics for evaluation of scholarly performance in the social sciences and humanities in Taiwan. *Proceedings of the American Society for Information Science and Technology, 49*, 1–7. https://doi.org/10.1002/meet.14504901060

Tijssen, R. J. W., Leeuw, J. D. & van Raan, A. F. J. (1988). A method for mapping bibliometric relations based on field-classifications and citations of articles. In L. Egghe & R. Rousseau (Eds), *Informetrics 87/88: Select Proceedings of the First International Conference on Bibliometrics and Theoretical Aspects of Information Retrieval, Diepenbeek, Belgium, 25–28 August 1987* (pp. 279–292). Amsterdam: Elsevier Science Publishers.

Tijssen, R. J. W. & van Raan, A. F. J. (1989). Mapping co-word structures: A comparison of multidimensional scaling and Leximappe. *Scientometrics, 15*, 283–295. https://doi.org/10.1007/BF02017203

Tomov, D. T. & Mutafov, G. H. (1996). Comparative indicators of interdisciplinarity in modern science. *Scientometrics, 37*, 267–278. https://doi.org/10.1007/BF02093624

Tsai, C.-C., Corley, E. A. & Bozeman, B. (2016). Collaboration experiences across scientific disciplines and cohorts. *Scientometrics, 108*, 505–529. https://doi.org/10.1007/s11192-016-1997-z

Tsay, M.-y. (2015). Citation type analysis for social science literature in Taiwan. In A. A. Salah, Y. T. A. A. A. Salah, C. Sugimoto & U. Al (Eds), *Proceedings of ISSI 2015 Istanbul: 15th International Society of Scientometrics and Informetrics Conference* (pp. 117–128). Istanbul: Boğaziçi University Printhouse.

van Leeuwen, T. (2004). Descriptive versus evaluative bibliometrics. In H. F. Moed, W. Glänzel & U. Schmoch (Eds), *Quantitative Science and Technology Research: The Use of Publication and Patent Statistics in Studies of S&T Systems* (pp. 373–388). New York: Kluwer Academic Publishers. https://doi.org/10.1007/1-4020-2755-9_17

van Raan, A. F. J. (1997). Scientometrics: State-of-the-art. *Scientometrics, 38*, 205–218. https://doi.org/10.1007/BF02461131

van Raan, A. F. J. (2004). Measuring science: Capita selecta of current main issues. In H. F. Moed, W. Glänzel & U. Schmoch (Eds), *Quantitative Science and Technology Research: The Use of Publication and Patent Statistics in Studies of S&T Systems* (pp. 19–50). New York: Kluwer Academic Publishers. https://doi.org/10.1007/1-4020-2755-9_2

Visser, M., Eck, Nees Jan v. & Waltman, L. (2019). Large-scale comparison of bibliographic data sources: Web of Science, Scopus, Dimensions, and Crossref. In G. Catalano, C. Daraio, M. Gregori, H. F. Moed & G. Ruocco (Eds), *Proceedings of the 17th Conference of the International Society for Scientometrics and Informetrics* (Vol. II, pp. 2358–2369). Rome: Edizioni Efesto.

Wagner, C. S. & Leydesdorff, L. (2005). Network structure, self-organization, and the growth of international collaboration in science. *Research Policy, 34*, 1608–1618. https://doi.org/10.1016/j.respol.2005.08.002

Wallin, J. A. (2005). Bibliometric methods: Pitfalls and possibilities. *Basic & Clinical Pharmacology & Toxicology, 97*, 261–275. DOI: 10.1111/j.1742-7843.2005.pto_139.x

Waltman, L. & Eck, N. J. v. (2012). The inconsistency of the *h*-index. *Journal of the American Society for Information Science and Technology, 63*, 406–415. DOI: 10.1002/asi.21678

Wang, L., Wang, X. & Philipsen, N. J. (2017). Network structure of scientific collaborations between China and the EU member states. *Scientometrics, 113*(2), 765–781. https://doi.org/10.1007/s11192-017-2488-6

Xia, X., Wang, Z., Wu, Y., Ruan, L. & Wang, L. (2014). Country of authorship and collaboration affect citations of articles by South and East Asian authors in agronomy journals: A case study of China, Japan, and India. *Serials Review, 40*, 118–122. https://doi.org/10.1080/00987913.2014.929610

Xian, H. & Madhavan, K. (2014). Anatomy of scholarly collaboration in engineering education: A big-data bibliometric analysis. *Journal of Engineering Education, 103*, 486–514. https://doi.org/10.1002/jee.20052

Yoshikane, F. & Kageura, K. (2004). Comparative analysis of coauthorship networks of different domains: The growth and change of networks. *Scientometrics, 60*, 433–444. https://doi.org/10.1023/B:SCIE.0000034385.05897.46

3

SCIENTOMETRICS IN THE HUMANITIES AND SOCIAL SCIENCES

Introduction

Scientometrics has been conceived to be a study of visible patterns of scholarly activities (Weingart, 2015). The activities pertain to research conducted at the individual, institutional and/or national levels. In the early years, scientometric studies were mostly performed within the sphere of the natural sciences (Sivertsen, 2009). Primarily because of the reasons related to its origin and its usefulness in studying science, scientometrics was popular with studies on science. The sources of data available then for scientometric research were mostly about science publications and references in science. Inquiries were, by and large, centred on the quality of articles conducting citation analyses in science.

Largely, scientometric studies were conducted by social scientists. Social scientists undertook the research enterprise to study what science is, what scientists produce and how it is communicated. They had, for a while, shied away from looking inwardly into what they and their peers have been doing in their own disciplines and fields of study. In studying the disciplines in the HSS, social scientists had not been as enthusiastic as they were in investigating science disciplines. Their own disciplines and subject areas were omitted conveniently or unintentionally. The scenario changed when there was enough data, mainly in the form of citation indexes, to study the HSS. Gradually, but steadily, the interests of social scientists in the HSS began to grow. The HSS began to be promising for scientometric analysis, but this was not as easy as studying science. Barriers had to be overcome when the HSS became the subject of scientometrics. This chapter examines such barriers and how they can be tackled, and looks into the potential of scientometric research in the HSS.

Potential for scientometrics in the humanities and social sciences

Focusing on science and non-science

The history of the ties between scientometrics and the HSS goes well back to the origin of the former. A revisit to the definitions of scientometrics makes clear the inseparability of science and non-science in scientometrics studies. In the definition of Vassily Nalimov (Nalimov and Mulchenko, 1969) who coined the term, scientometrics is about the scientific activities undertaken in *all disciplines* [emphasis added]. Science or the HSS were not segregated or excluded from this definition. Braun et al. (1987: 2) define how "scientometrics analyses the quantitative aspects of the generation, propagation, and utilisation of *scientific information* [emphasis added] in order to contribute to a better understanding of the mechanism of *scientific research activities* [emphasis added]". Both definitions do not suggest only science, but also encompass all disciplines and scientific research.

Péter Vinkler (2010) clarifies in his book the increased use of the method for the study of subjects other than sciences. Scientometrics may belong to the discipline of the science of science, but it should not be regarded as a field above other scientific fields. Scientometrics is not the science of sciences, but a science on science for science (Vinkler, 2010). Vinkler believes that scientometrics covers different areas and aspects of all sciences, and that its laws, rules and relationships cannot be considered as being as exact or as hard as those of the natural sciences, but also not as lenient (soft) as those of social science disciplines. In Lancho-Barrantes' (2018) discussion of the purpose of scientometrics to engage the techniques of the metrics for the evaluation of science, the term science refers to both the pure and the social sciences. Scientometrics interacts with several disciplinary fields such as sociology, information sciences, the philosophy of science, the history of science and economics, and it integrates different approaches into it (Ivancheva, 2008).

Scientometric research relies largely on economic, sociological and informational models for the interpretation of data (Gómez-Morales, 2015). Both the founders and practitioners of scientometrics acknowledged this a long time ago. In March 1962, Garfield wrote to Price on the former's *Science Citation Index* (*SCI*) project: "[T]here appears to be a great deal that the sociologist can do with citation indexes" (Garfield, 1988: 349). Garfield was encouraged to launch citation indexes for the HSS, paving the way for scientometric research in these areas. This was more than an indication of the larger role sociologists and social scientists can play in the use of citation indexes for the study of the larger scientific domain. Garfield visualised the contribution that sociologists could make in this field with citation indexes. Admittedly, scientometrics is more associated with the sociological understanding of science as a social system (Gómez-Morales, 2015). It is the job of sociologists and not scientists to study the context of science, production and dissemination of knowledge and their impacts that scientists produce. At the same time, it is also the responsibility of sociologists and social scientists to see how their own disciplines are performing from time to time, and what is happening in their own academic backyards.

Scientometric approaches are as relevant and appropriate for all subjects as for science. Having evolved into an acceptable set of methods, its applications have been extended and expanded beyond the boundaries of science. It is now a complete disciplinary field with clearly defined subjects of research and specific sets of methods and techniques (Ivancheva, 2008). It is one of the truly interdisciplinary research fields to extend to all scientific fields (Glänzel and Schoepflin, 1994), by consolidating components from mathematics, social sciences, natural sciences, engineering and life sciences (Glänzel, 2003). Known as a growing interdisciplinary area, it is integrated at the subject matter and an applicational area for different contributing fields. Physicists, chemists, mathematicians, medical scientists, sociologists, psychologists, philosophers and historians are in the community of scientometricians (Glänzel and Schoepflin, 1994). The disciplines in the HSS cannot therefore be omitted from the scope of scientometric applications.

Studying HSS

Although scientometrics came into being and was primarily embraced as an effective tool for the study of science, its applications in other realms of scientific knowledge were experimented with. Scientometrics progressively gained attention in the study of disciplines and subjects in the HSS. Social scientists opened the disciplines in the HSS to scientometric analysis. The interest in scientometrics for the study of the HSS kept growing among scholars, who found the databases useful to conduct serious examination of their own fields and disciplines. This is evident from the studies which have appeared over the last few years.[1] A number of scientometric studies on the HSS have already been published, which will be examined in the next chapter. Many scientometric analyses were performed by researchers in their own fields, subjects and disciplines. From such studies it is obvious that scientometric studies were not strictly confined to the domains of natural, life and engineering sciences alone. Now scientometrics has a broader scope than what was confined to the realm of science. Currently, both science and the HSS fall under the territory of scientometrics.

What can be studied in the HSS?

What should be studied in the HSS with scientometric methods? Reviewing studies in the HSS, Hammarfelt (2016) reports that it largely depends on the availability of data sources, and the content, availability and coverage of these data sources. This is the main precondition for any scientometric study. The purpose of scientometrics in the HSS, as Hammarfelt declares, is not to study databases or coverage, but to advance knowledge of the communication structures in research. Scientometric studies therefore have a more serious role to play in making sense of the nature of research and publications in the HSS.

Revealed in studies varying from the histories of scientific discoveries to patterns of development in the HSS (Gómez-Morales, 2015) is the effectiveness of scientometrics that cuts across disciplinary boundaries. The method assists both the HSS and social scientists in many different ways. More and more, social scientists realise the application of scientometrics in subfields such as the history of science, science, technology and innovation policy, the sociology of science, scientific collaboration, networks and several other fields that can benefit the disciplines in the HSS. Interest in scientometric research of the HSS among policy makers and analysts is showing growing trends (Leydesdorff et al., 2011). With the growth in the community of scholars attracted to citation data, scientometrics became increasingly important to the field of policy and to the development of scholarship (Franssen and Wouters, 2019). The social policy perspective of scientometrics, according to Franssen and Wouters (2019), invoked the types of data and the methods used. Scientometrics was used to examine the publication profiles in the humanities of universities and countries, rather than the examination of references in the science domain (Franssen and Wouters, 2019). Since then, a whole range of topics pertaining to the HSS has been explored using scientometric methods. The review of scientometric studies in the humanities by Franssen and Wouters supports the analytical value of different approaches to the study of the HSS.

In line with the argument of Franssen and Wouters (2019), scientometrics has a particular representation of science influencing the science system. In the case of the HSS, it offers insights into the broader questions of the HSS. Franssen and Wouters look at the ways in which the social and cognitive structures and publication practices have been studied in scientometrics, which go beyond mere research evaluation. A variety of studies with scientometrics has occurred in the HSS. Franssen and Wouters classify them into three types of studies. They pertain to scientometric indicators and research evaluation; science-mapping techniques that study the cognitive structure of the humanities; and the use of national and regional databases for the study of publication practices. The following chapter provides more details of these.

Knowledge production is a broad area of inquiry that can be studied with the help of scientometric data. Through the analysis of published documents, the production of knowledge and its characteristics of trends around subject areas, years, institutions and countries are made known. Production trends in disciplines and subjects are an area of attraction for social scientists. This is the opportunity for social scientists to emulate those applications in science and adapt them to the HSS.

Unfortunately, the potential of scientometrics in the study of the HSS disciplines was not explored or exploited in the way it warranted. Such efforts would have been advantageous for both the community of social scientists and their own respective disciplines. This is an opportunity that might have added value and direction to the development of many of the HSS disciplines.

One of the aims of scientometric studies is to map the development (or decline) of disciplines, and to find the way they have been advancing or declining. Research evaluation is another core area. Quite a few studies of research evaluation and assessment have been concluded in the HSS. The most important purpose of research evaluation is to make research accountable and attain knowledge of research strengths and weaknesses, at both individual and institutional levels (Loprieno et al., 2016).

Mapping of the disciplines in the HSS has a host of purposes. Mapping enables researchers to identify the strengths and weaknesses of the subfields within disciplines. It can retrieve information, help in understanding the dynamics of disciplines and inform policies regarding the allocation of resources and funding (Leydesdorff and Milojević, 2015). Apart from these, mapping unravels the trajectories of development of a discipline or field of study, and how it has functioned over the years. The technique of mapping can further improve knowledge of the emerging or declining fields within disciplines.

Being a major area in scientometrics, productivity can be analysed in any given research field in the HSS (Andrés, 2009). In this area, scholars view scientometrics as a tool, for measuring social sciences as it does for the natural, biomedical and engineering sciences (Godin, 2002). Important and well-recognised reports in the social sciences, such as the *World Social Science Report* (ISSC, IDS and UNESCO, 2016), make use of scientometric data as an integral component for in-depth analysis.

The social connections in HSS publications bring in qualitative nuances to scientometrics (Sula, 2012). Social data on connections between authors seen in publications bridges the gaps in scientometric studies. Sula (2012) argues that the methods of analysis for the visualisation developed for the humanities can also be applied to sciences. It means scientometric studies in the HSS can contribute to science, as the latter contributes to the HSS.

Scientometrics allows for the analysis of a range of areas and topics that are close to the heart of social scientists. The use and applications of scientometrics for the HSS are now beyond question and are widely acknowledged and recognised. The foci of scientometrics shifted as scientometricians began to look inwardly into their own disciplines and subjects.

Although this is the case and practice, social scientists have not regarded this as an equally attractive research area such as the more popular areas of gender studies, development, globalisation or climate change. The applications of scientometrics to the HSS, however, lag behind science disciplines. This continues to be the situation when the HSS are advancing and are opening new avenues for social scientists to study communication and knowledge in the HSS.

Metrics and indicators for HSS studies

New metrics for scientometric applications to the HSS have been developed. For the measurement of research performance, the metrics developed by Tang et al. (2013) take into account the diverse publication formats prevalent in the HSS. By developing a group of alternative metrics with a distinctive domestic nucleus, Tang et al. make a contribution to the development of scientometric studies of disciplines, subjects and research areas in the HSS. Discipline-specific refinements have also been undertaken in the HSS. Aistleitner et al. (2018) investigated the impact of quantitative indices in scientometrics for the development of the discipline of economics.

Indicators are required for the research assessment pertaining to individual scholars and are crucial tools in the scientometric toolkit. In an overview of author-level indicators, Wildgaard et al. (2014) presented a few which can be applied to research assessment. They are categorised as indicators of publication count, output on the level of the researcher and journal, the effect of the output, such as citations and citations normalised to field or research work, and the rank of the individual's work and impact over time. Taking into account the structure of research activities and research output in the HSS, Moed et al. (2002) developed a set of indicators for research assessment. These indicators were drawn from the field of law in the humanities. They refer to the unit of analysis (publications, journal and individual), publication output classification (18 types),[2] publication output in journals, ranking of journals and weighting of publication output and international orientation. Moed et al. followed a seven-step procedure for calculating the indicators which have application not only for the field of law, but also for other subjects in the HSS.[3]

Despite the evidence of citation studies, the number of studies using scientometric methods are limited in the humanities (Ardanuy, 2013). The interests of social scientists who applied scientometrics for their research were mostly centred on publications, patents, citations and indexes. This shows that the HSS face some challenges with scientometrics which need to be investigated.

Challenges for the humanities and social sciences

Issues of challenges

When a closer look is taken at those scientometric studies published in prominent scientometrics journals and the books that are based on scientometric data, it is apparent that only a minute fraction of the studies have been dealt with in the HSS disciplines. This lack of progress and achievement in the HSS is mainly due to:

- The general perception that scientometrics is relevant and useful only for the study of the sociology of science. The same perception continues to influence social scientists and deter them from undertaking scientometric studies. This validates the limited number of scientometric studies in the HSS.
- Scientometrics was extensively used for mapping science and allied matters of science, with the exception of studies conducted in library and information science. Scholars from other disciplines in the HSS do not therefore consider it as their area of study.
- The topics covered by scholars that applied the method gave the impression that it requires quantitative and statistical skills on the part of students and researchers, which intimidated those who wanted to enter the area. In a large number of scientometric studies, both in science and the HSS, this has been the case. The use of complex mathematical formulae and models in the analysis is

apparently a deterrent for scholars in the HSS, who are not trained in mathematics or statistics.

- Hardly any attempts to generate qualitative data for scientometric studies have been undertaken. The potential for the mining of qualitative data from databases has not been fully exploited or explored. This prevented researchers who are trained and experienced in dealing with qualitative data from doing scientometric research.
- As in the study of science and how science is produced, disseminated and used by the scientific community, scholars had preconceptions about the scope of scientometrics for other areas and topics of research.
- The natural sciences are well represented in the databases while the HSS are only partially covered (Bornmann and Leydesdorff, 2014). This limits the scope of analyses for the HSS.
- The differing publication practices in the HSS in general, and within the HSS in particular, continue to be a challenge for scientometric studies. The value of books and monographs in the HSS is higher in comparison to science. The plurality of disciplines makes the HSS a heterogeneous entity. Scholars (Ball, 2018, for instance) have grouped the disciplines describing their differentiating publication cultures. The characteristic nature of disciplines makes scientometric applications in the HSS a complex exercise.

Along with these, as Akker (2016) summarises, difficulties are noticed in some broad areas. Firstly, there are substantial differences in scientific practice between several disciplines within the humanities, which has consequences for quality indicators required for research assessment. Secondly, the rotation time of articles and books will have a longer effect than it has in the natural and life sciences. They are cited even after many years of their publication. Thirdly, there are different goals and products of research in different areas of the humanities. The goals and products are not limited to articles and books, but include the construction of large databases, exhibitions and materials in archives, museums and libraries. Fourthly, there are differences in the publication channels. In some disciplines within the humanities, books are still the main or even the only accepted way for the dissemination of knowledge, whereas it is articles in some other disciplines. Fifthly, the issue of language in which the knowledge is produced in the humanities has a strong nationalistic concern. Finally, there is the level or lack of organisation that exists in the humanities.

Given these challenges, it is important to elaborate on some of them. They mainly refer to the unique publication practices in the HSS and the coverage of available data. Both the publication practices and the coverage are interrelated and influence each other.

Publication practices in the HSS

Publication practices are neither stable nor do they remain unchanged for longer periods of time in the history of disciplines. Disciplines, regardless of their connections in the natural sciences or the HSS, are prone to this feature. As far as the challenges of scientometrics for the study of the HSS are concerned, some characteristic features that distinguish the publication practices within the HSS from science are to be taken into account. It is typical of the HSS to have their knowledge produced not only in the form of journal publications, but also in other publication formats. Publication practices in the HSS are therefore characterised by a pluralism of publication types (Thiedig, 2019). The importance of scholarly books in the HSS is indisputable (Giménez-Toledo et al., 2016). Kyvik (2003) states that the distribution of articles, books and reports in the HSS is even, while the distribution is skewed in favour of journal articles for the natural sciences. Monographs and books are not as common in the natural sciences as in the HSS. They, however, play significant roles in research communication in the HSS (Al et al., 2006; Kousha and Thelwall, 2009). In contrast to science, journal articles account for 45 to 70 per cent of the research output in the social sciences and 20 to 35 per cent in the humanities (Archambault and Gagné, 2004). Scholarly monographs in the HSS are still being relied upon rather heavily as both primary and secondary sources of information (Thompson, 2002). For scholars in science, journal articles account for the major source of scientific knowledge (Larivière et al., 2006). Publication practices in the HSS therefore differ considerably from those in scientific, technical and biomedical fields (Verleysen and Weeren, 2016). While launching the *Social Sciences Citation Index* (*SSCI*) in 1973, Garfield was aware that monographs and books are important in the social sciences and that they are cited more than journal publications (Garfield, 1976).

Kulczycki and Korytkowski (2019), in their recent research, showed that monographs are the key publication channel for the HSS and that scholars who publish monographs also publish journal articles.[4] Ferrara et al. (2019) studied changes in publication practices in the HSS in Italian academia, following the introduction of procedures to improve the quality of research in Italy. While the share of journal articles had increased progressively over the period of analysis, monographs retained their central position in all areas of the HSS (Ferrara et al., 2019).

The influence of the type of communication of publications through non-indexed journals, books, proceedings, reports and those in the electronic media, is not less important (van Raan, 1997). Compared to science, disciplines in the HSS produce a significant volume of knowledge in non-journal publication formats. In the study of the HSS, books and national and non-scholarly publications are included along with journal publications (Hicks, 2004). Such non-scholarly and non-journal literature is more important in research communication in the HSS than in science (Larivière et al., 2006). This makes the information retrieval for the HSS different from science, because the publication documents required differ due to the fundamental nature of scientific research in the HSS (Garfield, 1980). This is a challenge, as the coverage of the knowledge that can be studied by employing scientometric methods is not comprehensive.

The dominance of monographs and books in the HSS is as evident as the change in the publication behaviour of scholars in the HSS. In a study of the publication behaviour of faculty members in Norwegian universities, Kyvik (2003) identified the changes in publications. He found that publishing is now more directed towards international audiences; publications in international journals are becoming the dominant type of publication; coauthorship is more common; and the output per academic is increasing. These findings resonate with the selection of the database, as international publications are becoming common and, hence, the databases that have such publications are useful, as well as the type of analysis required regarding the output of researchers and coauthorships, which is a proxy to scientific collaboration.

Corresponding to the spread and acceptance of scientometrics in the social sciences, no standard methods are used in the humanities (Ball, 2018). Ball (2018) argues for the development of indicators that can study research and knowledge in the HSS. This should be done in collaboration with individual disciplines, taking into account the specific publication habits of disciplines.

In reviewing the current practices of scientometrics in the HSS, Archambault and Gagné (2004) warn that scientometrics should be used with care and caution in the HSS. If applied without proper consideration of the differing disciplinary cultures and publication practices, scientometric studies can affect the validity of the findings derived (Cox et al., 2019). Further, the heterogeneity of research topics, methods and paradigms in the HSS make it more difficult to set a standard criterion for research assessment (Ochsner et al., 2017) for the entire HSS. These issues and singularities of the HSS as a whole, and the individual disciplines separately, are to be accommodated in research assessment and evaluation.

Ball (2018) compared the features of the publications with regard to thematic orientation, publication language, place and kind of publication, target group and authorship (sole author/coauthor). Nederhof (2006) agrees with Ball on several of the above points that differentiate between disciplines in science and in the HSS and that pose challenges to scientometrics in the HSS. Thematic orientation is international for the natural sciences, but the HSS are more strongly influenced by the results of national academic output. Publications in the natural sciences, by and large, aim at the international specialist audience, while in the HSS they are aimed at national specialist academics and audiences. Research in the HSS is more embedded in the social context and is therefore more influenced by national trends and national policy concerns (Hicks, 2004). Scholars in the HSS manifest more nationally oriented publication and citation behaviour (Pajić et al., 2019). A considerable amount of research output in some of the fields in the HSS is primarily oriented at the national, regional and local public, and they are mainly published in regional/national outlets. Given the limited use of scholarly journals in the HSS, the national-oriented ones are more relevant than the international-oriented ones (Hammarfelt, 2016; Toledo, 2016). The HSS deal with phenomena that are specific to a geographical and social context. It is also to be accepted that the regional or national research in the HSS conducted in one country may not necessarily evoke the same appeal in other countries (Nederhof, 2006; Toledo, 2016).

In the HSS, the publication language is often the language of the country where the paper is produced, whereas for the natural sciences it is largely English. Social sciences have more competing paradigms than those in the natural sciences (Katz, 1999). Authorship patterns in the HSS do not seem to be similar to those in science. Coauthorship is more common in the natural sciences than in the HSS. It is often single authors who produce works in the HSS, while several authors are engaged in the production of publications in the natural sciences. This is evident in the analysis of the number of authors in the HSS and science.

Disparities between science and the HSS apart, variations among the disciplines within the HSS are not inconsequential. Differences in the research practices and, as a consequence, publication practices, owing to technical developments in the disciplines and specialities within the HSS are to be considered (Hammarfelt, 2016) and taken seriously in scientometric studies. Van Leeuwen's study (2006) indicated differences in publication cultures across disciplines, while a strong resemblance of publication behaviour is exhibited across nations within the social sciences. Engels et al. (2012) collected the changing publication patterns in the HSS in Flanders in Belgium. The dominance of journal articles in the humanities is less pronounced, whereas book publications accounted for more than one quarter of the outputs in art, history, linguistics, literature and theology (Engels et al., 2012). In an earlier study, Lindholm-Romantschuk and Warner (1996) examined the role of monographs versus journal articles in philosophy, economics and sociology. The study unfolded the relative impact of monographs and journal articles, which varied across disciplines. Monographs in philosophy seemed to have a relatively more significant impact than in sociology and economics. Philosophy monographs tended to receive more citations than monographs in sociology and economics.

Publication patterns within the HSS were released in an extensive study (n=74,022) carried out by a group of scholars (11 authors from eight countries) (Kulczycki et al., 2018). Pertaining to the period of 2011–2014, the study considered publications in all the HSS disciplines, including economics and business, law, philosophy and theology. The publications were collected from the Czech Republic, Denmark, Finland, Flanders (Belgium), Norway, Poland, Slovakia and Slovenia. In order to address the coverage of the prominent databases, they used peer-reviewed publications registered in national databases, and extended the study to journal articles, monographs, edited volumes, chapters in books and conference proceedings in English, the local language and other languages. Dissimilar publication patterns among fields, within fields and among countries were obvious. While the publication patterns in the HSS are stable and similar in the West European and Nordic countries, considerable changes have been observed in the Central and Eastern European countries. Both the share of the articles and the share of publications in English have been on the rise in these countries. The types of publication (monographs and journal publications) had variance across disciplines and countries. It is critical to note, as shown in this study, that in non-English speaking countries, the patterns in the language and types of publication were determined by the norms, culture and expectations of the disciplines. Decisively, publications were determined by each country's cultural and historic heritage, science policy and research incentives.

Citation practices are also part of publication practices which differ, depending on the understanding of the role of research (Nelhans, 2014). Differences between science and the HSS are obvious in citation behaviour. The use of references in the humanities is generally connected to the concepts of originality, intellectual organisation and searching and writing (Hellqvist, 2010). A stronger citation pattern of books and book chapters is reported in the HSS than in science (Toledo, 2016). The proportion of citations of journals in the HSS journal literature is half of that which was observed in the natural sciences and engineering (Larivière et al., 2006). The analysis of Yates and Chapman (2007) on references in three journals in the field of communication showed dissimilar reference practices. Differences in citation patterns of books and journals across disciplines within the social sciences have also been reported (Jokić et al., 2019). Remarkably, the age span of citations in the HSS is longer and broader, because scholars use resources that cover a broad age span (Hammarfelt, 2016).

The above features of the HSS cannot be neglected while planning to design scientometric studies of the disciplines in the HSS.

Sources of data and their coverage

While the interest and benefits of scientometrics in the HSS continue to swell, the source of data and its coverage for scientometric studies remain a challenge. The presence of databases that cover publications in the HSS allows scholars to understand their history and practice, and what they mean (Zuccala, 2016). A major issue is the lack of adequate coverage of publication forms in major databases. Adequate coverage is important to gain accurate knowledge about productivity and publication behaviour in diverse disciplines in the HSS (Ochsner et al., 2017). A large share of sources and targets, as Glänzel and Chi (2019) note, is located outside the research community. The main constraint of scientometrics to the study of the HSS disciplines is in the non-comprehensive nature of data. Therefore, it is worthwhile to explore some of the characteristics of the HSS that suit scientometric studies. It is in this context that the recent debate on the representation of the HSS in scientometrics becomes timeous.

Coverage in terms of language should also be taken into account in choosing a database for scientometric studies. Social scientists, compared to natural scientists, write and read fewer foreign language or foreign journals (Hicks, 2004). Not all journals are covered by the indexes, only the major ones on the market. In comparing the coverage of non-English publications in WoS and Scopus, Olga and Akoev (2019) realised that the share of English publications has increased over the non-English publications in the databases. Therefore, the role of non-English publications in scientometric uses, like research evaluation, in non-English speaking countries is underestimated. This is more so when research in the HSS is assessed only by publications in WoSCC and Scopus (Olga and Akoev, 2019). It should be remembered that, as van Leeuwen et al. (2001) remind, the impact indicators of research activities at the level of institution or country depend on the inclusion and exclusion of publications written in languages other than English.

Over the last few years, several databases available on the market have begun to collect and store data on publications in the HSS. The arrival of citation indexes exclusively meant for the HSS disciplines made scientometric studies possible in an area which has been hitherto underdeveloped. With the launch of the *SSCI* and the *Arts & Humanities Citation Index* (*A&HCI*) in 1973 and 1978 respectively, the attention of scholars to conduct scientometric studies in the HSS grew considerably. Following this, two other major databases, Scopus and Google Scholar, changed the situation with improved coverage of publications in the HSS. The increased availability of data on publications also made research on a variety of aspects of the HSS easier. The citation indexes that hold the HSS publications make the study of these subjects feasible. A strong empirical support for scientometric research on the HSS has thus been created. As a result of these, the scope and potential of scientometrics in the HSS is now more extensive than before.

A study of 13 databases in the HSS (Sīle et al., 2018) found that differences in national databases are related to differences in country-specific arrangements. The comprehensiveness of the databases was examined, to search for national databases in the HSS; the type of research output and the timespan of research output included in the databases; and how the data of research output was collected. Most of the examined databases collected research outputs from any academic discipline, and some were broader than others in their coverage.

The coverage limitations of the HSS publications in prominent databases are being addressed with the use of additional databases that have a local, national and regional content and focus. Such nationally oriented producers and consumers of social science are quite evident in scientometric studies (Hicks, 2004). While the same scientometric methods may be appealing to the study of disciplines in the HSS, some extensions are necessary. Nederhof (2006) recommends that, in addition to WoS data sources, additional sources such as non-WoS serials, monographs, contributions to edited volumes, reports and publications directed at a non-scholarly public must be sought. This is in agreement with the characteristic publishing practices prevalent in the HSS. The emergence of local and national data repositories in Asia, Latin America and the Nordic countries has facilitated in addressing the challenge of coverage of the HSS literature.[5]

Since inadequate coverage has implications for both research assessment and funding, some countries have taken the lead in improving the coverage of databases to make research assessment more reliable. In Europe, measures were taken to overcome the problems of underrepresentation of journals and publications of the HSS in databases. As part of the European Scoping Project, the possibility of developing a database capturing a full range of publications from the HSS was sought (Martin et al., 2010). The purpose of the project is also to develop performance measures for assessing research quality and impact in the HSS.[6] Countries such as Belgium, Croatia, Czech Republic, Denmark, Estonia, Finland, Hungry, Norway, Portugal, Slovenia, Spain and Sweden have already established or are in the process of building national current research information systems. These systems have bibliographic metadata of their scholarly publication outputs, and are endowed with the potential for a more comprehensive coverage of scholarly publications in the HSS (Sivertsen, 2014).

On the part of scholars, strategies to overcome the difficulties raised by coverage inadequacies are being tried. In this bid, they focus on the characteristic features of individual disciplines; supplement major databases with other local, national or regional ones; and include all types of publication other than journal articles.

More options to overcome this limitation to the study of subjects and disciplines in the HSS that are not the primary concerns of citation indexes are available. One option is to choose a journal or a group of journals that are core to the publications in a discipline or subject area. Such studies illustrate the focus of the journals, themes that the journal has carried in its lifetime and the trends. Information about papers can be collected from selected journals and processed using a data management program to create independent datasets meant for small studies for specific purposes. This is dealt with in Chapter 5.

Social scientists have experimented with ways of combining scientometric material with sociological data. This is an added opportunity for the HSS to address the limitations that can lead to more meaningful contributions in the HSS. By combining methods and data, Mählck (2001) created socio-bibliometric maps that explained the department-based research activity. When writing the history of science and social science disciplines, Sooryamoorthy (2015, 2016) and Ruggunan and Sooryamoorthy (2019) adopted scientometric methods alongside historical and sociological methods. Similarly, studies have incorporated additional information such as the gender and race of authors drawn from other sources. Such innovative studies make a strong case for the use of scientometrics in the HSS. More than ever, there is a need to focus on the disciplines and subjects in the HSS.

Nevertheless, these problems and challenges should not be used as an excuse for not attempting scientometric studies in the HSS. The issues pertaining to the HSS create an awareness of the need to address and solve these problems in order to conduct studies that will lead to valid findings and outcomes. In the next chapter, the scope of scientometrics in terms of its applications is detailed, with the support of specific cases and examples.

Notes

1 To cite some of the studies: Aistleitner et al., 2018; Anglada-Tort and Sanfilippo, 2019; Ardanuy, 2013; Benckendorff and Zehrer, 2013; Bornmann et al., 2016; Chandra, 2017; Chi, 2012; Diem and Wolter, 2013; Franssen and Wouters, 2019; Gumpenberger et al., 2016; Guns et al., 2018; Hammarfelt, 2016; Hellqvist, 2010; Ho and Ho, 2015; Huang and Chang, 2008; Köseoglu et al., 2016; Kulczycki et al., 2018; Leydesdorff et al., 2011; Merigó et al., 2018; Mosbah-Natanson and Gingras, 2014; Nederhof et al., 2010; Nwagwu and Egbon, 2011; Ochsner et al., 2017; Ossenblok et al., 2014; Ossenblok and Engels, 2015; Prins et al., 2016; Ruggunan and Sooryamoorthy, 2019; Serenko et al., 2009; Sin, 2011; Sivertsen and Larsen, 2012; Sooryamoorthy, 2016, 2017a, 2017b; Sula, 2012; Tang et al., 2013; Toledo, 2016; Tripathi and Babbar, 2018; van Leeuwen, 2013; Verleysen and Engels, 2014; Verleysen and Weeren, 2016; White et al., 2016; Zhu et al., 2013; Zuccala, 2016; and Zyoud et al., 2018.

2 The classification includes: 1. Books published by a single author; 2. Published PhD thesis; 3. Book published as a coauthor; 4. Unpublished PhD thesis; 5. Substantial scholarly contribution published in accepted scholarly journals, anniversary volumes, seminar reports, collective works; 6. Edited book or collective work; 7. Published integral contribution to international conferences; 8. Published abstract of lecture in international conferences; 9. Published integral contribution to national conferences; 10. Published abstract of lecture at national conferences; 11. Scholarly contribution of limited size; 12. Teaching course notes; 13. Scholarly edition of codes of law, jurisdiction volumes, bibliographies; 14. Research report for the scholarly community; 15. Internal research report or commissioned work; 16. Published inaugural or valedictory lecture; 17. Other publications such as an introduction, editorial contribution, letter to the editor, commemorative article, correction, book review; and 18. Judicial publications for a wide audience.

3 The steps are: 1. Collect raw data in electronic form for every scholar on publication output; 2. Reduplicate entries; 3. Identify book publications; 4. Identify PhD theses and avoid double-counting as a book; 5. Identify all publications that exceed five pages; 6. Add up the number of items 3 to 5, determining the "raw" number of "core" publications; and 7. Calculate the weighted number of core publications, by weighting journal articles with the journal weights and book publications by a factor obtained by dividing the number of pages by 16, the median page length of a substantial contribution (Moed et al., 2002).

4 The study covered publications of 67,415 Polish researchers that included 30,185 monographs and 638,779 articles published during 2013–2016.

5 Some of the databases that could address insufficient coverage of data in the HSS are: the Chinese Science Citation Database; China Scientific and Technical Papers and Citations; the Russian Index of Science Citation; the SciELO from Brazil (with a focus on national journals and including research from Latin America, Spain, Portugal, the Caribbean and South Africa); Lattes Platform (for Latin American countries); Sucupira Platform (Brazil); ASEAN Citation Index (representing scientific research in Brunei Darussalam, Cambodia, Indonesia, Lao People's Democratic Republic, Malaysia, Myanmar, the Philippines, Singapore, Thailand and Vietnam); the Taiwan Humanities Citation Index and the *Taiwan Social Sciences Citation Index* (for Taiwan); the Academic Citation Index (Taiwanese publications in the social sciences); the Citation Database for Japanese Papers; the Serbian Social Science Citation Index; and the Norwegian Scientific Index (used in Norway).

6 The project identified the obstacles and the scope for developing a database for the HSS that would cover all relevant publications for the HSS, including articles in international and national journals, book chapters, monographs, books and other non-published and non-textual research inputs.

References

Aistleitner, M., Kapeller, J. & Steinerberger, S. (2018). The power of scientometrics and the development of economics. *Journal of Economic Issues*, *52*, 816–834. DOI: 10.1080/00213624.2018.1498721

Akker, W. v. d. (2016). Yes we should; research assessment in the humanities. In M. Ochsner, S. E. Hug & H.-D. Daniel (Eds), *Research Assessment in the Humanities Towards Criteria and Procedures* (pp. 23–30). Zürich: Springer. https://doi.org/10.1007/978-3-319-29016-4_3

Al, U., Sahiner, M. & Tonta, Y. (2006). Arts and humanities literature: Bibliometric characteristics of contributions by Turkish authors. *Journal of the American Society for Information Science and Technology*, *57*, 1011–1022. https://doi.org/10.1002/asi.20366

Andrés, A. (2009). *Measuring Academic Research: How to Undertake a Bibliometric Study.* Oxford: Chandos Publishing.

Anglada-Tort, M. & Sanfilippo, K. R. M. (2019). Visualizing music psychology: A biblio-metric analysis of psychology of music, music perception, and musicae scientiae from 1973 to 2017. *Music & Science, 2,* DOI: 10.1177/2059204318811786.

Archambault, É. & Gagné, É.V. (2004). *The Use of Bibliometrics in the Social Sciences and Humanities.* Montreal: Social Sciences and Humanities Research Council of Canada (SSHRCC). www.science-metrix.com/pdf/SM_2004_008_SSHRC_Bibliometrics_Social_Science.pdf

Ardanuy, J. (2013). Sixty years of citation analysis studies in the humanities (1951–2010). *Journal of the American Society for Information Science and Technology, 64,* 1751–1755. https://doi.org/10.1002/asi.22835

Ball, R. (2018). *An Introduction to Bibliometrics: New Development and Trends.* Cambridge, MA: Chandos Publishing.

Benckendorff, P. & Zehrer, A. (2013). A network of analysis of tourism research. *Annals of Tourism Research, 13,* 121–149. https://doi.org/10.1016/j.annals.2013.04.005

Bornmann, L. & Leydesdorff, L. (2014). Scientometrics in a changing research landscape. *EMBO reports, 15,* 1228–1232. DOI: 10.15252/embr.201439608

Bornmann, L., Thor, A., Marx, W. & Schier, H. (2016). The application of bibliometrics to research evaluation in the humanities and social sciences: An exploratory study using normalized Google Scholar data for the publications of a research institute. *Journal of the American Society for Information Science and Technology, 67,* 2778–2789. https://doi.org/10.1002/asi.23627

Braun, T., Bujdosò, E. & Schubert, A. (1987). *Literature of Analytical Chemistry: A Scientometric Evaluation.* Boca Raton, FL: CRC Press.

Chandra, Y. (2017). Mapping the evolution of entrepreneurship as a field of research (1990±2013): A scientometric analysis. *PLOS One, 13.* https://doi.org/10.1371/journal.pone.0190228.

Chi, P.-S. (2012). Bibliometric characteristics of political science research in Germany. *Proceedings of the American Society for Information Science and Technology, 49,* 1–6. https://doi.org/10.1002/meet.14504901115

Cox, A., Gadd, E., Petersohn, S. & Sbaffi, L. (2019). Competencies for bibliometrics. *Journal of Librarianship and Information Sciences, 51,* 746–762. https://doi.org/10.1177/0961000617728111

Diem A. & Wolter, S. C. (2013). The use of bibliometrics to measure research performance in education sciences. *Research in Higher Education, 54,* 86–114. https://doi.org/10.1007/s11162-012-9264-5

Engels, T. C. E., Ossenblok, T. L. B. & Spruyt, E. H. J. (2012). Changing publication patterns in the Social Sciences and Humanities, 2000–2009. *Scientometrics, 93,* 373–390. https://doi.org/10.1007/s11192-012-0680-2

Ferrara, A., Nappi, C. A. & Pentassuglio, F. (2019). Changing publication practices in Italy: The case of Social Sciences and Humanities. In G. Catalano, C. Daraio, M. Gregori, H. F. Moed & G. Ruocco (Eds), *Proceedings of the 17th Conference of the International Society for Scientometrics and Informetrics* (Vol. II, pp. 1507–1518). Rome: Edizioni Efesto.

Franssen, T. & Wouters, P. (2019). Science and its significant other: Representing the human-ities in bibliometric scholarship. *Journal of the American Society for Information Science and Technology, 70,* 1124–1137. https://doi.org/10.1002/asi.24206

Garfield, E. (1976). On the literature of the social sciences and the usage and effectiveness of the *Social Sciences Citation Index. Current Contents, 34,* 5–10. www.garfield.library.upenn.edu/essays/v2p550y1974-76.pdf

Garfield, E. (1980). Is information retrieval in the arts and humanities inherently different from that in science? The effect that ISI's citation index for the arts and humanities is expected to have on future scholarship. *Library Quarterly, 50,* 40–57. DOI: 10.1086/629874

Garfield, E. (1988). Derek Price and the practical world of scientometrics. *Science, Technology & Human Values, 13*, 349–350. https://doi.org/10.1177/016224398801303-412

Giménez-Toledo, E., Mañana-Rodríguez, J., Engels, T. C. E., Ingwersen, P., Pölönen, J., Sivertsen, G., … Zuccala, A. A. (2016). Taking scholarly books into account: Current developments in five European countries. *Scientometrics, 105*, 685–699. https://doi.org/10.1007/s11192-016-1886-5

Glänzel, W. (2003). Bibliometrics as a research field: A course on theory and application of bibliometric indicators. Retrieved from http://yunus.hacettepe.edu.tr/~tonta/courses/spring2011/bby704/bibliometrics-as-a-research-field-Bib_Module_KUL.pdf 21 May 2019.

Glänzel, W. & Chi, P.-S. (2019). Research beyond scholarly communication – The big challenge of scientometrics 2.0. In G. Catalano, C. Daraio, M. Gregori, H. F. Moed & G. Ruocco (Eds), *Proceedings of the 17th Conference of the International Society for Scientometrics and Informetrics* (Vol. I, pp. 423–436). Rome: Edizioni Efesto.

Glänzel, W. & Schoepflin, U. (1994). Little scientometrics, big scientometrics … and beyond? *Scientometrics, 30*, 375–384. https://doi.org/10.1007/BF02018107

Godin, B. (2002). *The Social Sciences in Canada: What Can We Learn from Bibliometrics. Working Paper No. 1.* Quebec: INRS, University of Quebec. www.csiic.ca/PDF/CSIIC.pdf

Gómez-Morales, Y. J. (2015). Scientometrics. In G. Ritzer (Ed.), *The Blackwell Encyclopedia of Sociology*. London: John Wiley & Sons.

Gumpenberger, C., Sorz, J., Wieland, M. & Gorraiz, J. (2016). Humanities and social sciences in the bibliometric spotlight – Research output analysis at the University of Vienna and considerations for increasing visibility. *Research Evaluation, 25*, 271–278. https://doi.org/10.1093/reseval/rvw013

Guns, R., Sïle, L., Eykens, J., Verleysen, F. T. & Engels, T. C. E. (2018). A comparison of cognitive and organizational classification of publications in the social sciences and humanities. *Scientometrics, 116*, 1093–1111. https://doi.org/10.1007/s11192-018-2775-x

Hammarfelt, B. (2016). Beyond coverage: Toward a bibliometrics for the humanities. In M. Ochsner, S. E. Hug & H.-D. Daniel (Eds), *Research Assessment in the Humanities Towards Criteria and Procedures* (pp. 115–132). Zürich: Springer. https://doi.org/10.1007/978-3-319-29016-4_10

Hellqvist, B. (2010). Referencing in the humanities and its implications for citation analysis. *Journal of the American Society for Information Science and Technology, 61*, 310–318. https://doi.org/10.1002/asi.21256

Hicks, D. (2004). The four literatures of social science. In H. F. Moed, W. Glänzel & U. Schmoch (Eds), *Handbook of Quantitative Science and Technology Research: The Use of Publication and Patent Statistics in Studies of S&T Systems* (pp. 476–496). Dordrecht, The Netherlands: Kluwer Academic Publishers. https://doi.org/10.1007/1-4020-2755-9_22

Ho, H.-C. & Ho, Y.-S. (2015). Publications in dance field in Arts & Humanities Citation Index: A bibliometric analysis. *Scientometrics, 105*, 1031–1040. https://doi.org/10.1007/s11192-015-1716-1

Huang, M.-h. & Chang, Y.-w. (2008). Characteristics of research output in social sciences and humanities: From a research evaluation perspective. *Journal of the American Society for Information Science and Technology, 59*, 1819–1828. https://doi.org/10.1002/asi.20885

ISSC, IDS & UNESCO. (2016). *World Social Science Report 2016. Challenging Inequalities: Pathways to a Just World.* Paris: UNESCO Publishing. https://en.unesco.org/wssr2016

Ivancheva, L. (2008). Scientometrics today: A methodological overview. *Collnet Journal of Scientometrics and Information Management Decision, 2*, 47–56. https://doi.org/10.1080/09737766.2008.10700853

Jokić, M., Mervar, A. & Mateljan, S. (2019). Comparative analysis of book citations in social science journals by Central and Eastern European authors. *Scientometrics, 120*, 1005–1029. https://doi.org/10.1007/s11192-019-03176-y

Katz, J. S. (1999). *Bibliometric Indicators and the Social Sciences.* Brighton, UK: University of Sussex.

Köseoglu, M. A., Rahimi, R., Okumus, F. & Liu, J. (2016). Bibliometric studies in tourism. *Annals of Tourism Research, 61,* 180–198. https://doi.org/10.1016/j.annals.2016.10.006

Kousha, K. & Thelwall, M. (2009). Google book search: Citation analysis for social science and the humanities. *Journal of the American Society for Information Science and Technology, 60,* 1537–1549. https://doi.org/10.1002/asi.21085

Kulczycki, E., Engels, T. C. E., Pölönen, J., Bruun, K., Duškova, M., Guns, R., ... Zuccala, A. (2018). Publication patterns in the social sciences and humanities: Evidence from eight European countries. *Scientometrics, 116,* 463–486. https://doi.org/10.1007/s11192-018-2711-0

Kulczycki, E. & Korytkowski, P. (2019). What share of researchers publish monographs? In G. Catalano, C. Daraio, M. Gregori, H. F. Moed & G. Ruocco (Eds), *Proceedings of the 17th Conference of the International Society for Scientometrics and Informetrics* (Vol. I, pp. 179–184). Rome: Edizioni Efesto.

Kyvik, S. (2003). Changing trends in publishing behaviour among university faculty, 1980–2000. *Scientometrics, 58,* 35–48. https://doi.org/10.1023/A:1025475423482

Lancho-Barrantes, B. S. (2018). Knowledge distribution through the Web: The webometrics ranking. In F. J. Cantú-Ortiz (Ed.), *Research Analytics: Boosting University Productivity and Competitiveness through Scientometrics* (pp. 161–183). Boca Raton, FL: CRC Press. DOI: 10.1201/9781315155890-10

Larivière, V., Archambault, É., Gingras, Y. & Vignola-Gagné, É. (2006). The place of serials in referencing practices: Comparing natural sciences and engineering with social sciences and humanities. *Journal of the American Society for Information Science and Technology, 57,* 997–1004. https://doi.org/10.1002/asi.20349

Leydesdorff, L., Hammarfelt, B. & Salah, A. (2011). The structure of the Arts & Humanities Citation Index: A mapping on the basis of aggregated citations among 1,157 journals. *Journal of the American Society for Information Science and Technology, 62,* 2414–2426. https://doi.org/10.1002/asi.21636

Leydesdorff, L. & Milojević, S. (2015). Scientometrics. In M. Lynch (Ed.), *International Encyclopedia of Social and Behavioral Sciences*: Amsterdam: Elsevier.

Lindholm-Romantschuk, Y. & Warner, J. (1996). The role of monographs in scholarly communication: An empirical study of philosophy, sociology and economics. *Journal of Documentation, 52,* 389–404. https://doi.org/10.1108/eb026972

Loprieno, A., Werlen, R., Hasgall, A. & Bregy, J. (2016). The "Mesurer les Performances de la Recherche" project of the Rectors' Conference of the Swiss Universities (CRUS) and its further development. In M. Ochsner, S. E. Hug & H.-D. Daniel (Eds), *Research Assessment in the Humanities Towards Criteria and Procedures* (pp. 13–22). Zürich: Springer. https://doi.org/10.1007/978-3-319-29016-4_2

Mählck, P. (2001). Mapping gender differences in scientific careers in social and bibliometric space. *Science, Technology & Human Values, 26,* 167–190. https://doi.org/10.1177/016224390102600203

Martin, B., Tang, P., Morgan, M., Glänzel, W., Hornbostel, S., Lauer, G., ... Zic-Fuchs, M. (2010). *Towards a Bibliometric Database for the Social Sciences and Humanities – A European Scoping Project. A report produced for DFG, ESRC, AHRC, NWO, ANR and ESF.* Brighton, UK: Science and Technology Policy Research Unit. https://globalhighered.files.wordpress.com/2010/07/esf_report_final_100309.pdf

Merigó, J., Pedrycz, W., Weber, R. & Sotta, C. d. l. (2018). Fifty years of information sciences: A bibliometric overview. *Information Sciences, 432,* 245–268. https://doi.org/10.1016/j.ins.2017.11.054

Moed, H. F., Luwel, M. & Nederhof, A. J. (2002). Towards research performance in the humanities. *Library Trends, 50*, 498–520.

Mosbah-Natanson, S. & Gingras, Y. (2014). The globalization of social sciences? Evidence from a quantitative analysis of 30 years of production, collaboration and citations in the social sciences (1980–2009). *Current Sociology, 62*, 626–646. https://doi.org/10.1177/0011392113498866

Nalimov, V. V. & Mulchenko, Z. M. (1969). *Naukometriya. Izuchenie nauki kak informatsionnogo protsessa (Scientometrics. Study of Science as an Information Process).* Moscow: Nauka. English version*: Measurement of Science. Study of the Development of Science as an Information Process.* Washington, DC.: Foreign Technology Division, US Air Force Systems Command. Retrieved from http://garfield.library.upenn.edu/nalimovmeasurementofscience/chapter1.pdf. 20 March 2019.

Nederhof, A. J. (2006). Bibliometric monitoring of research performance in the social sciences and the humanities: A review. *Scientometrics, 66*, 81–100. https://doi.org/10.1007/s11192-006-0007-2

Nederhof, A. J., van Leeuwen, T. N. & Raan, A. F. J. v. (2010). Highly cited non-journal publications in political science, economics and psychology: A first exploration. *Scientometrics, 83*, 363–374. https://doi.org/10.1007/s11192-009-0086-y

Nelhans, G. (2014). Qualitative scientometrics. *Proceedings of the IATUL Conference*, Paper 6. http://docs.lib.purdue.edu/iatul/2014/plenaries/2016.

Nwagwu, W. & Egbon, O. (2011). Bibliometric analysis of Nigeria's social science and arts and humanities publications in Thomson Scientific databases. *The Electronic Library, 29*, 438–456. DOI: 10.1108/02640471111156722

Ochsner, M., Hug, S. & Galleron, I. (2017). The future of research assessment in the humanities: Bottom-up assessment procedures. *Palgrave Communications, 3*. DOI: 10.1057/palcomms.2017.20

Olga, M. & Akoev, M. (2019). Non-English language publications in citation indexes – Quantity and quality. In G. Catalano, C. Daraio, M. Gregori, H. F. Moed & G. Ruocco (Eds), *Proceedings of the 17th Conference of the International Society for Scientometrics and Informetrics* (Vol. I, pp. 35–46). Rome: Edizioni Efesto.

Ossenblok, T. L. B. & Engels, T. C. E. (2015). Edited books in the social sciences and humanities: Characteristics and collaboration analysis. *Scientometrics, 104*, 219–237. https://doi.org/10.1007/s11192-015-1544-3

Ossenblok, T. L. B., Verleysen, F. T. & Engels, T. C. E. (2014). Coauthorship of journal articles and book chapters in the social sciences and humanities (2000–2010). *Journal of the American Society for Information Science and Technology, 65*, 882–897. https://doi.org/10.1002/asi.23015

Pajić, D., Jevremov, T. & Škorić, M. (2019). Publication and citation patterns in the social sciences and humanities: A national perspective. *Canadian Journal of Sociology, 44*, 67–94. DOI: 10.29173/cjs29214

Prins, A. A. M., Costas, R., van Leeuwen, T. N. & Wouters, P. F. (2016). Using Google Scholar in research evaluation of humanities and social science programs: A comparison with Web of Science data. *Research Evaluation, 25*, 264–270. https://doi.org/10.1093/reseval/rvv049

Ruggunan, S. & Sooryamoorthy, R. (2019). *Management Studies in South Africa: Exploring the Trajectory in the Apartheid Era and Beyond.* Basel, Switzerland: Springer.

Serenko, A., Bontis, N. & Grant, J. (2009). A scientometric analysis of the proceedings of the McMaster World Congress on the Management of Intellectual Capital and Innovation for the 1996–2008 period. *Journal of Intellectual Capital, 10*, 8–21. https://doi.org/10.1108/14691930910922860

Sīle, L., Pölönen, J., Sivertsen, G., Guns, R., Engels, T. C. E., Arefiev, P., … Teitelbaum, R. (2018). Comprehensiveness of national bibliographic databases for social sciences and humanities: Findings from a European survey. *Research Evaluation*, *27*, 310–322. https://doi.org/10.1093/reseval/rvy016

Sin, S.-C. J. (2011). International coauthorship and citation impact: A bibliometric study of six LIS journals, 1980–2008. *Journal of the American Society for Information Science and Technology*, *62*, 1770–1783. https://doi.org/10.1002/asi.21572

Sivertsen, G. (2009). Publication patterns in all fields. In F. Aström, R. Danell, B. Larsen & J. W. Schneider (Eds), *Celebrating Scholarly Communication Studies: A Festschrift for Olle Persson at His 60th Birthday* (pp. 55–60). Leuven, Belgium: International Society for Scientometrics and Informetrics.

Sivertsen, G. (2014). Scholarly publication patterns in the social sciences and humanities and their coverage in Scopus and Web of Science. Paper presented at the Context Counts: Pathways to Master Big and Little Data, Leiden, The Netherlands.

Sivertsen, G. & Larsen, B. (2012). Comprehensive bibliographic coverage of the social sciences and humanities in a citation index: An empirical analysis of the potential. *Scientometrics*, *91*, 567–575. https://doi.org/10.1007/s11192-011-0615-3

Sooryamoorthy, R. (2015). *Transforming Science in South Africa: Development, Collaboration and Productivity*. Hampshire and New York: Palgrave Macmillan.

Sooryamoorthy, R. (2016). *Sociology in South Africa: Colonial, Apartheid and Democratic Forms*. Hampshire and New York: Palgrave Macmillan.

Sooryamoorthy, R. (2017a). Do types of collaboration change citation? A scientometric analysis of social science publications in South Africa. *Scientometrics*, *111*, 379–400. https://doi.org/10.1007/s11192-017-2265-6

Sooryamoorthy, R. (2017b). Sociological research and collaboration in South Africa: Past and present. *International Sociology*, *32*, 567–586. https://doi.org/10.1177/0268580917725235

Sula, C. A. (2012). Visualizing social connections in the humanities: Beyond bibliometrics. *Bulletin of the American Society for Information Science and Technology*, *38*, 31–35. DOI: 10.1002/bult.2012.1720380409

Tang, M.-c., Wang, C.-m., Chen, K.-h. & Hsiang, J. (2013). Exploring alternative cyberbibliometrics for evaluation of scholarly performance in the social sciences and humanities in Taiwan. *Proceedings of the American Society for Information Science and Technology*, *49*, 1–7. https://doi.org/10.1002/meet.14504901060

Thiedig, C. (2019). The social sciences and their publishers: Publication, reception and changing meaning of German monographs. In G. Catalano, C. Daraio, M. Gregori, H. F. Moed & G. Ruocco (Eds), *Proceedings of the 17th Conference of the International Society for Scientometrics and Informetrics* (Vol. II, pp. 2014–2019). Rome: Edizioni Efesto.

Thompson, J. W. (2002). The death of the scholarly monograph in the Humanities? Citation patterns in literary scholarship. *Libri*, *52*, 121–136. https://doi.org/10.1515/LIBR.2002.121

Toledo, E. G. (2016). Assessment of journal & book publishers in the humanities and social sciences in Spain. In M. Ochsner, S. E. Hug & H.-D. Daniel (Eds), *Research Assessment in the Humanities Towards Criteria and Procedures* (pp. 91–102). Zürich: Springer. https://doi.org/10.1007/978-3-319-29016-4_8

Tripathi, M. & Babbar, S. K. P. (2018). Bibliometrics of social science and humanities research in India. *Current Science*, *114*, 2240–2247. DOI: 10.18520/cs/v114/i11/2240–2247

van Leeuwen, T. (2006). The application of bibliometric analyses in the evaluation of social science research: Who benefits from it, and why it is still feasible. *Scientometrics*, *66*, 133–154. DOI: 10.1007/s11192-006-0010-7 van Leeuwen, T. (2013). Publication trends in social psychology journals: A long-term biblio-metric analysis. *European Journal of Social Psychology*, *43*, 9–11. https://doi.org/10.1002/ejsp.1933

van Leeuwen, T. N., Moed, H. F., Tijssen, R. J. W., Visser, M. S. & van Raan, A. F. J. (2001). Language biases in the coverage of the Science Citation Index and its consequences for international comparisons of national research performance. *Scientometrics*, 51, 335–346. https://doi.org/10.1023/A:1010549719484

van Raan, A. F. J. (1997). Scientometrics: State-of-the-art. *Scientometrics*, *38*, 205–218. https://doi.org/10.1007/BF02461131

Verleysen, F. T. & Engels, T. C. E. (2014). Internationalization of peer reviewed and non-peer reviewed book publications in the social sciences and humanities. *Scientometrics*, *101*, 1431–1444. https://doi.org/10.1007/s11192-014-1267-x

Verleysen, F. T. & Weeren, A. (2016). Mapping diversity of publication patterns in the social sciences and humanities: An approach making use of fuzzy cluster analysis. *Journal of Data and Information Sciences*, *1*, 33–59. DOI: 10.20309/jdis.201624

Vinkler, P. (2010). *The Evaluation of Research by Scientometric Indicators*. Oxford: Chandos Publishing.

Weingart, S. B. (2015). Finding the history and philosophy of science. *Erkenntnis*, *80*, 201–213. https://doi.org/10.1007/s10670-014-9621-1

White, I., Guldiken, O., Hemphill, T., He, W. & Khoobdeh, M. S. (2016). Trends in inter-national strategic management research, from 2000 to 2013: Text mining and biblio-metric analyses. *Management International Review*, *56*, 35–65. https://doi.org/10.1007/s11575-015-0260-9

Wildgaard, L., Schneider, J. W. & Larsen, B. (2014). A review of the characteristics of 108 author-level bibliometric indicators. *Scientometrics*, *101*, 125–158. https://doi.org/10.1007/s11192-014-1423-3

Yates, S. D. & Chapman, K. (2007). An examination of the use of monographs in the com-munication journal literature. *Behavioral & Social Sciences Librarian*, *26*, 39–51. https://doi.org/10.1300/J103v26n01_03

Zhu, W., Zhu, J. G. & Guan, J. (2013). A bibliometric study of service innovation research: Based on complex network analysis. *Scientometrics*, *94*, 1195–1216. https://doi.org/10.1007/s11192-012-0888-1

Zuccala, A. (2016). Inciting the metric oriented humanist: Teaching bibliometrics in a fac-ulty of humanities. *Education for Information*, *32*, 149–164. DOI: 10.3233/EFI-150969

Zyoud, S. e. H., Sweileh, W. M., Awang, R. & Al-Jabi, S. W. (2018). Global trends in research related to social media in psychology: Mapping and bibliometric analysis. *International Journal of Mental Health System*, *12*. https://doi.org/10.1186/s13033-13018-10182-13036

4

CASES OF SCIENTOMETRIC STUDIES IN THE HUMANITIES AND SOCIAL SCIENCES

Introduction

Having learned the distinctiveness of the HSS disciplines and their potential and challenges for scientometric studies, it is worth dissecting some significant scientometric studies that have been successfully concluded in various disciplines in the HSS. This examination is expected to reveal different uses of scientometrics in the HSS, the context of each study, databases used, methodologies followed, data analysis procedures employed and the usefulness of the findings.

The cases and examples in the chapter point to the extent of the possibilities for scientometric research in the HSS, and open a new horizon of creative applications. In the analysis of cases, a framework of pertinent aspects is adopted. This takes into account the title and year of publications, title of the journals, chapters or books where applicable, authors and their institutional affiliation, data sources, research objectives/questions, methods, analysis and findings. This format is to assist readers to problematise their scientometric studies along these lines, or to deviate from these by charting new areas of scientometric inquiry. One incidental benefit that can be derived from these cases is that the prominent journals that publish scientometric studies are made known. The coauthorship and collaboration features hidden under the number of authors and their institutional affiliations are easy to find from the format adopted for the cases presented in the chapter. The disciplinary background of authors is meant to show who is studying which disciplines. Sources of data indicate various applications while methodologies expose possibilities.

Altogether, the chapter gathers 64 cases and examples of representative scientometric studies published in disciplines in the HSS. Some of the cases and examples are novel attempts, unravelling the concealed dimensions of scientific communication. The creative use of publications, some of which are quite unconventional in scientometric studies, is obvious in the HSS. This contributes to a deeper understanding of the disciplines and subjects in the HSS. From these cases and examples, one would be able to discern the entire range of the uses of publication data for the purpose of the study of the disciplines and subject areas in the HSS. The scope of scientometrics is expanded when new variables are created and are used innovatively. This is clear from the examination of the cases and examples.

Research evaluation and impact

Research evaluation and research impact are central to scientometric studies. Researchers use publications in the HSS to study scientific research and impact. Liu et al. (2012) examined research patterns in globalisation research, as seen in the publication records stored in WoS databases of the *Science Citation Index* (*SCI*), *Social Sciences Citation Index* (*SSCI*) and online library catalogues such as WorldCat and the US Library of Congress. The nature and characteristics of the research output in the HSS and their impact on the research evaluation were explored by Huang and Chang (2008). Phelan (2000) focused on the sociological research in Australia by studying publications in sociology as a method in research evaluation.

Larsen et al. (2008) combined publications in science and the HSS. The study is a comparison of publications and impact in China, Europe, India, Japan and the USA. Employing a basic analysis procedure of counting publications and citations, Larsen et al. were able to determine the shifting trends in scientific production in the selected countries and Europe (Case 4.1). A combination of databases, including the *SSCI* and the *Arts & Humanities Citation Index* (*A&HCI*), was adopted as data sources. The authors come from science and education disciplines.

CASE 4.1 PUBLICATION PRODUCTIVITY AND IMPACT

Title of the Publication: Scientific output and impact: Relative positions of China, Europe, India, Japan and the USA

Title of the Journal: *Collnet Journal of Scientometrics and Information Management*

Authors: Larsen, P. O., Maye, I. & Ins, M. v.

Institutional Affiliation: Center for Science and Technology (CEST), Switzerland; Secretariat for Education and Research (SER), Switzerland; and the Institute for Research Information and Quality Assurance, Germany

Year of Publication: 2008

Data Source: WoS datasets of the *SCI, SSCI* and *A&HCI*

Research Objective/Questions: Examine publication outputs, world shares of production and their impact in China, Europe, India, Japan and the USA.

Data, Methods and Analysis: The selected WoS datasets for the period of 1981–2004 were used. The share of the total production of publications (articles, notes, reviews and letters) in the selected countries and a union of countries showed decreasing, stagnating and increasing trends. The reasons, trends and impact are discussed. The study reveals that production of publications has been stagnating or gradually decreasing in Europe, the USA and Japan while it has been growing fast in China. Regarding the impact, the USA continued to maintain its lead. Findings are presented graphically as well.

Bjarnason and Sigfusdottir (2002) studied their own discipline of sociology. Using scientometric data, they researched the productivity and impact of the publications of the sociology faculty in 16 departments in five Nordic countries (Sweden, Finland, Norway, Denmark and Iceland). Appraising publications over a period of 20 years between 1981 and 2000, the study examined the publication productivity of sociology departments and citations. Journal articles indexed in the *SSCI* and *Sociological Abstracts* formed the data basis. Multilevel modelling performed in the study indicated that the departmental productivity predicts citations. Some of the key findings, as shown in Case 4.2, reveal the relationship between the publication patterns of the faculty and departments, and how this mutually contributed to the trajectories of both the faculty and the departments. The significance of this study is its ability to predict the connection between productivity and impact.

CASE 4.2 PRODUCTION AND IMPACT OF SOCIOLOGY DEPARTMENTS IN NORDIC COUNTRIES

Title of the Publication: Nordic impact: Article productivity and citation patterns in sixteen Nordic sociology departments

Title of the Journal: *Acta Sociologica*

Authors: Bjarnason, T. & Sigfusdottir, I. D.

Institutional Affiliation: Department of Sociology, University of Albany, USA; Icelandic Centre for Social Research and Analysis. Iceland; and Pennsylvania State University, USA

Year of Publication: 2002

Data Source: Multiple sources of the *SSCI*, *Journal Citation Reports* and *Sociological Abstracts*

Research Objective/Questions: Examine the production of publications by sociology faculties in Nordic countries, the impact of publications, and variations in productivity and impact across departments.

Data, Methods and Analysis: The analysis pertains to 1,205 articles in 329 journals. Impact factors for journals, authors and departments are measured.

The study presents the outlets in which the sociology faculty published their research (general/speciality sociology journals, national-language/English-language journals) and citation impacts of these publications. The analysis suggests that the sum of journal impact factors (JIF) can be used as a measure of the impact of individual scholars and sociology departments. Publications of authors who published more in high-impact journals received more citations. The same also applied to departments. Findings show that the publication patterns of individuals and departments raise the profile of faculty members and the departments alike.

A similar study assessed research productivity and impact. The study of Aaltojärvi et al. (2008) is similar in the context of the same department, but at a different time with dissimilar objectives and data sources. They used data from Google Scholar (GS) to examine the publication productivity, impact and web visibility of the faculty members in 16 Nordic sociology departments (Case 4.3). The focus is more on the web visibility of the researchers and the impact of their publications. While scientometric studies rely mainly on citation indexes of WoS or Scopus due to their known advantages over other databases, scholars adopt GS to compare aspects like web visibility with other databases. This case shows the application of the method on the information stored in GS, which is believed to be limited for detailed scientometric analyses.

CASE 4.3 PUBLICATION PRODUCTIVITY, WEB VISIBILITY AND CITATIONS OF SOCIOLOGISTS

Title of publication: Scientific productivity, web visibility and citation patterns in sixteen Nordic sociology departments

Title of the Journal: *Acta Sociologica*

Authors: Aaltojärvi, I., Arminen, I., Auranen, O. & Pasanen, H.-M.

Institutional Affiliation: Department of Sociology and Social Psychology, and the Institute for Social Research, University of Tampere, Finland

Year of Publication: 2008

Data Source: GS

Research Objectives/Questions: This is a study of sociology, sociologists and sociology departments by sociologists to find out the web visibility of Nordic (Sweden, Finland, Norway, Iceland and Denmark) sociology, the patterns of publication productivity and impact of Nordic sociology, and to compare the coverage of web visibility and publication productivity in GS and the *SSCI* of WoS. Factors that are relevant in the web visibility and impact are also investigated.

Data, Methods and Analysis: Data, drawn from GS, is about the faculty in the 16 sociology departments in Nordic countries. The names and positions of faculty members were gathered from the departmental web pages, which were then searched on GS for information on publications and citations. Publication details of 353 faculty members were analysed. The number of citations of the most cited publications were collected. Descriptive statistics on the faculties and their publications and citations are presented. Multilevel regression tests were applied to examine the relationship between citations and gender, position (rank), type of publication, country and the age of publications. Correlation between web visibility and the position of the faculty is found to be strong in the data, and web visibility, citations and positions of the faculty tend to influence each other. This multilevel study shows the visibility of individual scholars, departments and countries.

The research by Bornmann et al. (2016) is a typical evaluation study of a research unit in the sphere of the HSS (Case 4.4). Using GS data, publications produced by the researchers in the institute were assessed for their impact by measuring citations, which were normalised for analysis. As the study suggests, research evaluation using scientometrics can be performed not only for individuals, countries, disciplines or subjects, but also for research institutions.

CASE 4.4 RESEARCH EVALUATION IN THE HUMANITIES AND SOCIAL SCIENCES

Title of the Publication: The application of bibliometrics to research evaluation in the humanities and social sciences: An exploratory study using normalized Google Scholar data for the publications of a research institute

Title of the Journal: *Journal of the American Society for Information Science and Technology*

Authors: Bornmann, L., Thor, A., Marx, W. & Schier, H.

Institutional Affiliation: Division for Science and Innovation Studies, Max Planck Society, Germany; University of Applied Sciences for Telecommunications Leipzig, Germany; and Information Service, Max Planck Institute for Solid State Research, Germany

Year of Publication: 2016

Data Source: GS, WoSCC and Scopus

Research Objective/Questions: This study of research evaluation in the HSS in a research institute was carried out by scholars from science, innovation and information fields. The focus is on the citation impact of the publications of the chosen institute by developing procedures for analysis. It undertakes a normalisation of citation impact. The convergent validity of GS data is compared with that of WoSCC and Scopus.

Data, Methods and Analysis: The publications of the institute in 2009 were searched from GS ($n=212$). The selected publications included journal articles, book chapters, conference papers, books and PhD dissertations. The normalisation of citations was done for the journal publications, conference proceedings and book chapters. A major finding is that most of the publications that were analysed in GS earned more citations than in WoSCC.

Measuring research performance of research teams, Sandström and Besselaar (2019) combined scientometric data with primary survey data. Four aspects of research performance were assessed: productivity in terms of field-adjusted volume of publications (total production and per senior), and field-normalised citation scores (total and per senior). The characteristics such as team composition, dynamics and culture and dimensions related to gender diversity were investigated. The approach of the study extends beyond the conventional use of scientometric data combined with the use of primary data.

In an evaluation study, Gumpenberger et al. (2016) looked at the research output in the HSS. A selected university was the unit of investigation for a six-year period (Case 4.5). The disciplines included in the study were language and area studies, historical studies (archaeology, arts and history) and social sciences (political science, communication science and nursing science). Publication data was gathered from the *SSCI* and *A&HCI* databases of WoS. In addition to this data, data from interviews was supplemented. The evaluation of the research output of the university was assessed at different levels of the publication characteristics: types of research output, authorship (single/team), language (English/German), peer-reviewed (or not) and coverage of research output in databases. Gumpenberger et al. reported an increasing number of publications listed in prominent databases. The prominent features of the publications produced by the university in the HSS disciplines were compiled. The analysis further revealed the increasing international visibility of the research output of the university; growth in the production of output in quantity and quality; increasing international collaboration; the need for strengthening and expanding peer review; the need for metadata of non-English publications for increased visibility; and the growing number of publications in English. These are the necessary parameters for a study on research evaluation.

CASE 4.5 RESEARCH ASSESSMENT IN THE HUMANITIES AND SOCIAL SCIENCES

Title of the Publication: Humanities and social sciences in the bibliometric spotlight – Research output analysis at the University of Vienna and considerations for increasing visibility

Title of the Journal: *Research Evaluation*

Authors: Gumpenberger, C., Sorz, J., Wieland, M. & Gorraiz, J.

Institutional Affiliation: Library and Archive Services (Bibliometrics Department), University of Vienna, Austria

Year of Publication: 2016

Data Source: WoS

Research Objective/Questions: Conducted by scholars from the library and bibliometric fields, the study provides a longitudinal analysis of the research output in the HSS subjects. The focus is on the research productivity of a university for a specific period of six years. Authorship, collaboration and visibility were the key variables.

Data, Methods and Analysis: Data was sourced from the *A&HCI* and *SSCI* of WoS. The analysis is based on the total output of publications that included journal articles, books, book chapters, conference proceedings, letters, reviews, working papers and other contributions, the number of authors (sole- or group-authored) and selected languages. Publications were also analysed in terms of open access publications, digital object identifier (DOI) and collaboration. The findings are presented both descriptively and diagrammatically. Descriptive statistics have been used. The characteristics of the publication output of the scholars in the HSS in the University of Vienna are detailed. These characteristics are important in basic production trends. The analysis suggests that there has been an increase in international collaboration among the scholars at the university and an increase in the number of English-language publications, peer-reviewed publications and papers indexed in WoS databases. The social sciences adopt faster publication habits of the natural sciences than the humanities.

Investigating the places of publication of monographs, edited books and book chapters, Verleysen and Engels (2014) performed an analysis of research internationalisation in the HSS. As a research evaluation exercise, the study has an important value mainly due to its coverage of monographs, books and chapters, both peer-reviewed and non-peer reviewed. Internationalisation is usually examined in multinational coauthorships found in journal publications. The study proves that internationalisation in the HSS can be measured in book publications as well. The publications analysed in the study are those produced by scholars in Flemish universities between 2002 and 2011. As mentioned earlier, the centre of analysis is the publication barycentre, which is an indicator of internationalisation. It is evident from the analysis of the publication barycentres of books and chapters that the existing gap between peer-reviewed and non-peer reviewed publications in the HSS is widening.

The data from GS is not commonly used in scientometric studies, but Prins et al. (2016) found it worthwhile in the evaluation of academic programmes in universities (Case 4.6). Research programmes in three fields, namely, education, pedagogical sciences and anthropology in the Netherlands, were evaluated. With regard to the quality of the GS data and in comparison to WoSCC data there were obvious challenges. The major contribution of this study lies in bringing the issue of reliability of GS data to the fore and in demonstrating how it can be improved to suit scientometric analyses. Moreover, the study represents an example of a scientometric study that makes use of the less utilised databases in scientometrics.

CASE 4.6 RESEARCH EVALUATION OF HSS PROGRAMMES USING DATA FROM GOOGLE SCHOLAR

Title of the Publication: Using Google Scholar in research evaluation of humanities and social science programs: A comparison with Web of Science data

Title of the Journal: *Research Evaluation*

Authors: Prins, A. A. M., Costas, R., van Leeuwen, T. N. & Wouters, P. F.

Institutional Affiliation: Research Management, The Netherlands; and the CTWS Centre for Science and Technology Studies, Leiden University, The Netherlands

Year of Publication: 2016

Data Source: GS and WoS

Research Objective/Questions: Study the practical applications of GS-based metrics in the assessment of research programmes, issues of data quality and the comparison of results based on GS and WoSCC data.

Data, Methods and Analysis: Thirteen research programmes under the education and pedagogical sciences (Ed/Ped) and five programmes in anthropology in six universities in the Netherlands were evaluated. Sixty publications per programme were taken. Citation data was downloaded from GS, which was scrutinised for accuracy and then compared with WoSCC data. The cited output for the fields of Ed/Ped and anthropology differed: 90 and 58 per cent of journal publications, respectively. Anthropology had a higher percentage of cited output of books, volumes and chapters. Comparison of coverages of the publications in GS and WoS indicated large differences. In citations, higher values of counts are provided in GS rather than in WoS.

Evaluation studies are also carried out to assess groups of countries or regions. Erfanmanesh et al. (2019), in their study, involved institutions from 13 countries, which were part of the ASEAN university network. Three main research questions directed this study: publication productivity of the ASEAN university network; comparison between institutions in research excellence, innovation performance and research prominence; and the authorship patterns in relation to collaboration among the members of the network. The results demonstrate that Singaporean universities accounted for about one-third of the output for the chosen years, followed by four Malaysian universities. The evaluation was also undertaken around citations of the outputs that included the number of citations, mean citations per publication, cited rate, field-weighted citation impact and $h5$-index. In a similar way, research excellence of universities can be ascertained by studying the research preferences, as has been done by Lancho-Barrantes and Cantu-Ortiz (2019) in their analysis.

Individual research assessment has also been undertaken with scientometric methods. Mutz and Daniel (2019) did an individual research evaluation using the psychometric measurement approach to study the performance capacity of researchers. Based on scientometric data, they created a psychometric model to capture the scientific performance of researchers. With the help of WoSCC and GS data, Diem and Wolter (2013) assessed the research performance of individual scholars in the field of education. The authors sought to compare how much the rating for individual research performance is dependent upon the two databases. For the individual research assessment, they constructed quantitative (number of publications in the databases) and qualitative measures (citation count). Apart from these objectives, the study looked at the inter-individual differences in research output (Case 4.7).

CASE 4.7 ASSESSING RESEARCH PERFORMANCE OF INDIVIDUAL SCHOLARS

Title of the Publication: The use of bibliometrics to measure research performance in education sciences

Title of the Journal: *Research in Higher Education*

Authors: Diem, A. & Wolter, S. C.

Institutional Affiliation: Swiss Coordination Centre for Research in Education (SCCRE), Switzerland

Year of Publication: 2013

Data Source: *SSCI* and *A&HCI* of WoS and GS

Research Objective/Questions: Assess the research performance of individual scholars in the field of education sciences in Switzerland; test whether the assessment of individual research output varies according to the inclusiveness/exclusiveness of the datasets used in the scientific field; and examine the extent of cultural and language-driven differences in publication practices in research assessment.

Data, Methods and Analysis: The sample for the study was all professors (full, associate, titular and assistant) in the education sciences in seven Swiss universities. Scientometric data for the sample professors was collected from the *SSCI*, *A&HCI* and GS. Indicators such as the number of publications and the citation impact, which formed the dependent variables, were used for the individual assessment of the sample population. Explanatory variables included academic age (years after obtaining doctorate), biological age, professional category (tenured or not), gender and language region (French-speaking and others). Multivariate regression and correlation analyses were performed. Positive correlations existed across all the indicators of research performance in both datasets. Correlations between the output (number of publications) and the outcome (citations) or between quantity and quality were apparent. Explanatory models of variances in individual research performances were evident only in WoS datasets of the *SSCI* and *A&HCI* and not in GS.

As seen in the above cases and examples, research evaluation, one of the key applications of scientometric studies, can be achieved in many different ways in the HSS. Evaluation of research of a country, region, university, research institution, research unit, research team, department or of individual scholars is possible with scientometric data gathered from a variety of sources such as citation indexes, journals, monographs or conference proceedings. In the assessment of research performance, the unit of analysis is the production output, authorship (collaboration, for instance), type of document (journals, monographs, chapters, reviews or others) and/or impact (citations, *h*-index, etc.). The cases also suggest the usefulness and appropriateness of different data sources for research evaluation.

Mapping of disciplines and subjects in the HSS

As elaborated upon in the previous chapters, a major advantage of scientometrics is its ability to map and trace the development of a discipline by analysing publications in the discipline over a period of time. In mapping, the growth, stagnation or decline of a discipline in an institution, region, country or in the world can be explained. The changing foci of the discipline can also be traced through mapping.

Working on the measurement of the social sciences, Godin (2002) mapped the scientific production of Canadian researchers (1981–2000) in the social field. His examination centred on some basic mapping indicators relevant to the production of publications of Canadian researchers. Godin considered the volume of papers, geographical and sectoral origin of papers, collaboration and specialisation. In the end, the study proved to be a method which is as good a tool for measuring the social sciences as it is for science. The findings illustrate the scope of scientometrics in mapping social sciences in a prominent productive country like Canada. The data sourced from the *SSCI* contained publications of articles, review articles and notes. Godin adopted a comparative approach using the publications across all the provinces in the country. As for the origins of publications, the sectors of publications (universities, hospitals, government and business) were useful. Collaboration being a factor in the production of knowledge, Godin described inter-sector and international collaboration. In any mapping study, specialisation of subjects is a key element. The specialisation index of Canadian scholars has been calculated. Godin described the status of the social sciences in Canada using scientometric tools.

Research specialisation at the institutional level often appears in scientometric literature. Sorz et al. (2019) determined the research specialisations of a university, using key research areas of the faculty in the STEM (science, technology, engineering and mathematics) and HSS fields. The criteria they adopted were high publication activity, measured by the number of publications, number of publications in top journals and high citation impact. Affirmed in the study is that scientometric indicators are effective tools for locating research specialisations.

The structure of disciplines has also been at the centre of inquiry. Benckendorff and Zehrer (2013), using co-citation analysis, employed the method to the discipline of tourism. The networks of highly cited scholars and works in prominent journals in the discipline are revealed.

Big data, as described in Chapter 2, is the buzz word now. Singh et al. (2015) sought to study the research conducted on big data during a five-year period between 2010 and 2014. This mapping exercise validates the use of scientometric methods to map the trends in an emerging area of research. Both WoSCC and Scopus data have been used. Included among the variables were the research output and its growth rate, authorship, collaboration, contributors to big data by countries, institutions and individuals, and discipline. Network analysis and visualisation of data have been attempted. Studies of this nature are of benefit to scholars who can keep track of the developments occurring in their own fields and plan their future research accordingly. In knowledge production, novelty and originality are valued and welcomed. It is not hard to find scholars venturing into new areas of research. A scientific analysis of disciplinary and subject areas using scientometric methods can throw light on such emerging areas of research.

In examining world social sciences during the recent three decades, Mosbah-Natanson and Gingras (2014) made a daring effort in their large-scale research. In the theoretical context of centre-periphery, the study covers all the major regions in the world (Case 4.8). It portrays the history of the social sciences, looking specifically at internationalisation, collaboration and citations across the globe. In a previous paper, the same authors investigated the places where social sciences are produced (Gingras and Mosbah–Natanson, 2010). The world distribution of social science journals and publications, and the languages in which the publications are written, is explored further.

CASE 4.8 THE WORLD OF THE SOCIAL SCIENCES IN THE LAST THREE DECADES

Title of the Publication: The globalization of social sciences? Evidence from a quantitative analysis of 30 years of production, collaboration and citations in the social sciences (1980–2009)

Title of the Journal: *Current Sociology*

Authors: Mosbah-Natanson, S. & Gingras, Y

Institutional Affiliation: Paris-Sorbonne University Abu Dhabi, UAE; and the Department of History, University of Quebec in Montreal, Canada

Year of Publication: 2014

Data Source: *SSCI* of WoS

Research Objective/Questions: Study world social sciences over a 30-year period focusing on the evolution, production of papers, international collaboration and citation patterns across major regions.

Data, Methods and Analysis: Based on publication records, the study is driven by the centre-periphery model. The data is treated according to seven regions, namely Europe, North America, Latin America, Africa, Asia, Oceania and the Commonwealth of Independent States. The cartography of world social sciences is presented. The authors found that North America and Europe were the two predominant regions, accounting for about 90 per cent of the publications in the social sciences. An increase in the production of publications and in the share of publications in all regions during the period was noticed. Two peripheral regions, Asia and Latin America, had seen a substantial rise in the production of publications during the period of study. Most of the social sciences journals originated either in North America or in Europe. A general increase in the growth of inter-regional collaboration was reported. Western social scientists tended to associate with their national and regional counterparts rather than with their international counterparts. The two most cited regions for the social sciences are Europe and North America.

The following case (4.9) is a mapping study of the discipline of psychology. Zyoud et al. (2018) investigated the growth of publications, citations, collaboration and productivity, and the emerging areas of interest. Authors are affiliated to the medical and health sciences institutions, indicating their disciplinary backgrounds. Adopting the frequently used terms in the publications relating to social media (Facebook, Twitter, LinkedIn, Snapchat and Instagram), mapping was done in the field of psychology. By downloading data from WoS, the authors were able to track the growing trends in the publications in the chosen field of study in the discipline of psychology. Not only trends, but also other scientometric parameters were examined, uncovering disciplinary features and their interrelationships. Inferential statistical tests and visualisation techniques were adopted. As the authors indicated, this is a pioneering study aimed at gathering global trends in the field. The implications for setting the research agenda and its advancement are obvious. Scientometric method is a tool to map the literature on subjects and fields of research interests as well as to locate the gaps in research.

CASE 4.9 MAPPING RESEARCH IN PSYCHOLOGY

Title of the Publication: Global trends in research related to social media in psychology: Mapping and bibliometric analysis

Title of the Journal: *International Journal of Mental Health System*

Authors: Zyoud, S. e. H., Sweileh, W. M., Awang, R. & Al Jabi, S. W

Institutional Affiliation: College of Medicine and Health Sciences, An Najah National University, Palestine; and the WHO Collaborating Centre for Drug Information, Malaysia

Year of Publication: 2018

Data Source: WoSCC

Research Objective/Questions: Map the trends in the research literature related to the most used social network in the field of psychology. Examine the growth in the publications, citations, international collaboration, author productivity and emerging topics.

Data, Methods and Analysis: The data consists of 959 publications related to social media that are part of psychology for the period of 2004–2015, collected from WoSCC. These publications were then analysed for the year of publications, institutions of the authors, the country/territory of the authors, type of document, language, *h*-index, the impact factor, collaboration among authors and citations. The publications were sorted and analysed in terms of the most productive countries in the publications in the field; the most active journals in the publications; the most cited publications; and the top productive institutions in the said publications. The relationship between countries and institutions of authors was explored. Statistical tests revealed the correlation between the number of publications in the field of social media and the number of publications related to social media in psychology. The data was mapped using a visualisation software.

Publications in psychology in the Middle East countries formed the subject of analysis for Biglu et al. (2013). Data was gathered from SCImago. Inter-country variations in the production of publications in psychology in the region over a 15-year period ending in 2010 were found. Israel topped the list with 82 per cent of citable publications in psychology in the Middle East region. The remaining 18 per cent of the publications were shared by Iran, Kuwait, Lebanon, the UAE, Iraq, Jordan, Oman, Qatar, Saudi Arabia, Syria and Yemen. The importance of this brief study is in the presentation of a map of a discipline in a region of several countries that have not previously been studied.

Choosing a few core journals from Scopus, Anglada-Tort and Sanfilippo (2019) outlined the growth of publications, citations, productivity, collaboration and research trends in music psychology. The first of its kind in the study of music psychology, this scientometric analysis examined publications in 49 countries and traced their growth between 1973 and 2017. The growth in publications with temporal features across countries, citation trends, collaboration and productivity (using Lotka's law as a theoretical framework) make the study a worthwhile contribution in the mapping of the subject area. Visualised mapping is graphically presented.

A new multidisciplinary research field of the psychology of tourism was found in a scientometric study by Barrios et al. (2008), who tracked the evolution of a field of research by collecting the publications for the period of 1990–2005. In order to do this, Barrios et al. considered a set of indicators, namely the numbers of papers and authors, productivity by country and collaboration (Case 4.10). The interaction among these indicators offers insights into this young field of research. Like any emerging field, it is growing among a small community of individual scholars.

CASE 4.10 FINDING AN EMERGING FIELD OF THE PSYCHOLOGY OF TOURISM

Title of the Publication: A bibliometric study of psychological research on tourism

Title of the Journal: *Scientometrics*

Authors: Barrios, M., Borrego, A., Vilaginés, A., Ollé, C. & Somozab, M.

Institutional Affiliation: Department of Methodology of Behavioural Sciences, and the Department of Library and Information Science, University of Barcelona, Spain

Year of Publication: 2008

Data Source: *A&HCI* and *SSCI*

Research Objective/Questions: Trace the evolution of the research field, the psychology of tourism, between 1990 and 2005.

Data, Methods and Analysis: Publications (articles and reviews) were searched using the keywords of attitudes, behaviour, motivation, perception, psychology, satisfaction and tourism. The JIF in which the collected publications appeared was gathered along with the number of citations for the articles. Institutional collaboration was analysed for articles written by more than one author. The SPSS program was used for data analysis. Theoretically, the study rests on the laws of Price, Bradford and Lotka. Inferential statistics (logistic models) were employed. During the period of analysis, there was a significant growth in the number of publications in the area of the psychology of tourism. Increased coauthorships and collaboration among institutions within the country and different countries were reported.

Journals have strong empirical basis for scientometric analyses. Köseoglu et al. (2016), a team of researchers from four different institutions in four different countries, relied on specific journals in tourism and hospitality. As an example of a scientometric research, the study is based on the data gathered directly from journals, and not from citation indexes. Since the publications used in the study were actually scientometric studies, it is a scientometric study of scientometric studies in the applied field of tourism and hospitality. Again, this is about a study of a discipline by researchers from the same discipline who wanted to keep track of the development in their own area of specialisation. As detailed in Case 4.11, the research applies some new scientometric techniques for the study of a discipline. Compared to scientometric studies involving large sample sizes, it is a micro-level study consisting of a small sample. But, as pointed out, the research is based on data obtained from journals and not from citation indexes that hold large datasets, which is not very common. The methodological novelties of the study make it worth emulating in other fields in the HSS.

CASE 4.11 EMERGING THEMES AND RESEARCH IMPLICATIONS IN TOURISM

Title of the Publication: Bibliometric studies in tourism

Title of the Journal: *Annals of Tourism Research*

Authors: Köseoglu, M. A., Rahimi, R., Okumus, F. & Liu, J.

Institutional Affiliation: School of Hotel and Tourism Management, The Hong Kong Polytechnic University, Hong Kong; Faculty of Social Sciences, University of Wolverhampton, UK; Hospitality Services Department, University of Central Florida, USA; and Hospitality and Service Management Department, Sun Yat-Sen University, China

Year of Publication: 2016

Data Source: Five leading and high-impact tourism-focused and four hospitality-focused journals: *Annals of Tourism Research*; *Journal of Sustainable Tourism*; *Tourism Management*; *Journal of Travel Research*; *International Journal of Tourism Research*; *International Journal of Hospitality Management*; *Cornell Hospitality Quarterly*; *International Journal of Contemporary Hospitality Management*; and *Journal of Hospitality and Tourism Research*.

Research Objective/Questions: Examine the issues related to the field of tourism. The objectives are to find out the emerging themes and research implications of bibliometric structures, challenges and barriers to bibliometric studies, and to assess the impact of bibliometric studies on theory development and the future of research.

Data, Methods and Analysis: Data was drawn from nine prominent journals in tourism and hospitality studies. This is a bibliometric study of bibliometric studies conducted in the field of tourism. All articles of bibliometric studies (n=190) published in the selected journals from their first issue in 1988 to the end of 2015 were gathered. The articles were selected using a keyword search of the titles of the articles. A qualitative content analysis of articles was carried out for the type of bibliometric methods, time periods, types of sample, databases used, software used, themes, types of document and visualisation methods employed. By gathering a detailed list of themes of studies conducted in tourism research, the study found that studies on tourism have employed novel methods.

van Leeuwen (2013) examined the publication trends in social psychology journals, gathering information on length, citations and references over the period between 1991 and 2010. A gradual increase in the number of journals in social psychology in WoSCC and in the number of articles, particularly since 2004–2005, was observed. Noticeable changes in the typology of the page length of articles (declining number of shorter articles and increasing number of longer articles), decreasing number of US publications, increasing average number of references from 36 to 50, and the growing coverage of journal literature in the references were other key findings of this mapping research. The historical and future trends are portrayed with a few relevant variables. The course of developments within a subfield and in relation to the main field of psychology are apparent.

One important outcome of mapping, as seen in some of the cases described above, is that it can supply information regarding emerging areas, themes and issues relating to the field of study. Serenko et al. (2009) researched the emergence, evolution and identity of the young discipline of knowledge management and intellectual capital. In view of the non-availability of publications indexed in databases, they extended their scientometric analysis to the collection of conference proceedings over a period of 13 years. The method allowed them to assess the growth and evolution of this new discipline. The features of this young discipline have also been captured (Case 4.12).

CASE 4.12 SCIENTOMETRIC STUDY ON THE EVOLUTION OF A DISCIPLINE

Title of the Publication: A scientometric analysis of the proceedings of the McMaster World Congress on the Management of Intellectual Capital and Innovation for the 1996–2008 period

Title of the Journal: *Journal of Intellectual Capital*

Authors: Serenko, A., Bontis, N. & Grant, J.

Institutional Affiliation: Faculty of Business Administration, Lakehead University, Canada; and DeGroote School of Business, McMaster University, Canada

Year of Publication: 2009

Data Source: Conference proceedings of the McMaster World Congress on the Management of Intellectual Capital and Innovation for the period 1996–2008.

Research Objective/Questions: Investigate the state and evolution of a young discipline, knowledge management and intellectual capital. Identify authorship distribution, production across countries, institutions and individuals, publication frequency and the methodologies of the presentations. The questions the authors try to answer relate to the research output for the country, institution and individual.

Data, Methods and Analysis: A total of 399 papers published in the proceedings for the period were analysed. Lotka's law was applied to guide the inquiry. Data is descriptively analysed and presented in frequency tables. A growth in the discipline is revealed; half of the papers were single-authored; most of the papers were from the USA, Canada, the UK, Spain and Australia; three institutions were identified as the most productive institutions; the most productive scholars were found; and case studies were the most preferred methodology for the discipline.

Mapping a discipline should not be constrained by the language bias of readily available datasets that are comprised of predominantly English publications. In the study given in Case 4.13, the bias towards the English language in the citation indexes has been dealt with by taking data of publications from non-English languages. In order to perform a better analysis and to overcome the limitations of existing databases in the HSS, supplementary data should be used. Although such data is not comparable to that of the major databases, it complements major data sources. Chi's study (2012) of political science research in Germany is based on the dataset developed from the publications of political scientists in two departments. The data is combined with that obtained from WoSCC.

CASE 4.13 POLITICAL SCIENCE RESEARCH IN GERMANY

Title of the Publication: Bibliometric characteristics of political science research in Germany.

Title of the Journal: *Proceedings of the American Society for Information Science and Technology*

Author: Chi, P.-S.

Institutional Affiliation: Institute for Research Information and Quality Assurance, Berlin, Germany

Year of Publication: 2012

Data Source: Publications of two top-ranking departments of political science during 2003–2007, collected from two official websites, institutional repositories and the German Social Science Literature Information System. The references cited in the publications and citations were obtained from the WoS in-house database of the Competence Centre for Bibliometrics for the German Science System.

Research Objective/Questions: Publication behaviour of political scientists in Germany through the study of citation and reference features.

Data, Methods and Analysis: A sample of publications (*n*=1,018) of 33 professors from two selected institutions was analysed. The collected data was then verified by the professors. Publications included chapters in books, journal articles, conference papers, edited books, books, working papers, presentations, reports, lectures/speeches, discussion papers and newspaper/ magazine articles. Descriptive statistics and diagrammatic representation of data were applied. The study reported that most of the publications were in German (57%). About 40 per cent of the publications were in English and the rest in other languages. The difference in the type of document is recorded. Book chapters, edited books, books and non-peer reviewed publications were mostly in German rather than in English. Papers in English received more citations than those in German. About 53 per cent of the references cited in the sample publications were indexed in WoSCC. German political scientists published papers mainly in German and European journals.

Analyses of publications in a specific field of study such as that done by Caillods (2016) are another use of scientometrics. Such efforts keep subjects and subfields as the nucleus of the analysis. As seen in Case 4.14, Caillods focused on publications on the theme of inequalities and social justice to find the key features of the publications and the origin of the subject. Caillods shows how a theme in the social sciences can be subjected to a deeper analysis with a view to confining a subject within the regions of its production. This is an example of mapping a subject field using scientometric methods. In this mapping, the author found that publications in the subject field were contributed not only by social scientists, but also by other scientists. The findings reveal quite a few hitherto unknown features of the subject.

CASE 4.14 EXAMINING KNOWLEDGE DIVIDE IN THE AREA OF INEQUALITY AND SOCIAL JUSTICE IN THE HSS

Title of Chapter: Knowledge divides: Social science production on inequalities and social justice

Title of Book: *World Social Science Report 2016. Challenging Inequalities: Pathways to a Just World*

Author: Caillods, F.

Institutional Affiliation: Advisor, International Social Science Council (ISSC), Paris, France

Year of Publication: 2016

Data Source: WoS and Scopus

Research Objective/Questions: Examine researchers who contribute to the study of inequality and social justice; the domain and fields of researchers who study inequality and social justice; and the region and country of these researchers.

Data, Methods and Analysis: The analysis presents the data (1992–2013) according to the research questions. As regards the domains and fields in which publications on inequality and social justice were conducted, the study reports that a large number of publications emerged from the social, economic and behavioural sciences domain. An equally large number of publications came from the health sciences domain. The subfields of the domain in which the publications appeared have also been distinguished. Also, the trends and changes in the production of publications in the domains and fields are presented. About half of the publications were from North America, followed by Western Europe, Oceania, Sub-Saharan Africa, East and Southern Europe, and Latin America. A relatively higher volume of publications originated from South Africa (sixth largest producer of publications on the theme), than from France, the Netherlands or other European countries. African researchers from all over the continent worked in South Africa while Indian specialists were mostly based in the UK or in the USA.

Social work appears to be in the picture where scientometrics is concerned. With a view to better the scope of the study of social work, attempts have been made to build the foundation for future scientometric analyses. Keeping this in mind, Perron et al. (2017) provided an extensive list of social work journals and publications in these journals for a span of 25 years. It adds to the development of a set of tools and strategies that would serve scientometric research in social work. Like the authors, the study will attract scholars from the sample discipline to the understudied discipline of social work.

Focusing on the field of dance in the humanities, Ho and Ho (2015) conducted a study of their own field. The research outputs in dance published in the prominent drama journals were gathered from the *A&HCI*. The study accomplishes the objective of portraying the characteristics of journal publications (extracted from all the outputs) of a lesser known field in the humanities (Case 4.15). Based on the analysis of the data for two decades, the authors conclude that publications in the field have appeared mostly in a single journal; have fewer collaborations; received a low number of citations; and that ballet was the most popular topic of research. Emerging topics in dance were also summarised in this paper.

CASE 4.15 ANALYSIS OF JOURNAL PUBLICATIONS IN DRAMA

Title of the Publication: Publications in dance field in Arts & Humanities Citation Index: A bibliometric analysis

Title of the Journal: *Scientometrics*

Authors: Ho, H.-C. & Ho, Y.-S

Institutional Affiliation: College of Dance, Taipei National University of the Arts, Taiwan; and Trend Research Centre, Asia University, Taiwan

Year of Publication: 2015

Data Source: *A&HCI*

Research Objective/Questions: Analyse the publications in the dance field with a focus on the types of publication, language, output of publications, distribution of articles in journals, research activity, trends in research focus and authorship.

Data, Methods and Analysis: Publications for the period of 1994–2013 (n=28,307) in eight journals included in the *A&HCI* database under the category of dance were collected. The data was captured in an Excel spreadsheet. From the records of publications, author's name, title of the article and journal, year of publication, number of references cited in the articles, number of pages, keywords and citations were collected. Documents have 31 items ranging from articles to music scores. From these, 5,109 articles were separated. Descriptive statistics were applied to the analysis. The average number of authors per publication remained unchanged between the years of analysis; 95 per cent of the articles were single-authored; the average page length of article and the number of references fluctuated between the years; and the number of articles across the journals varied. Twenty most frequently used words showing the research foci were identified.

A scientometric study on the evolution of the field of entrepreneurship between 1990 and 2013 by Chandra (2017) supports the effectiveness of the method in understanding the history, evolution and emergence of new fields of knowledge. In this mapping study, combined methods such as topic mapping, author and journal co-citation, and visualisation were employed. Chandra was able to track the evolution of entrepreneurship research, its institutionalisation and legitimacy, and future opportunities. Approximately 46 pluralistic topics that existed in the history of entrepreneurship research, and that have been persistent across the years, and its interdisciplinarity character have also been revealed.

In an unusual study, Lewison et al. (2019) collected research literature on warfare and health. Investigating the effects of war on soldiers and civilians, they found that several countries were active in research on warfare and health. With the objective of tracing the development of Roman law from a historical perspective, Pölönen and Hammarfelt (2019) found data from GS useful. Publication records were gathered from GS that contained Roman law, including in the title of the publication and in the five languages of English, French, German, Italian and Spanish for the period between 1500 and 2016. The software Publish or Perish was utilised to collect the documents from GS. The data was systematically analysed to establish the number of publications and authors in the selected language groups; the average number of publications per author; and the concentrations of publications and citations. The growth in the field was estimated on the basis of the absolute and average number of publications and the number of authors involved in their production. In the researchers' view, GS is a capable tool for the historical analysis of scientometric data which is not available in other prominent databases. Nevertheless, Pölönen and Hammarfelt (2019) are open about the drawbacks of GS, such as the low quality of data and the difficulty in reproducing the results due to continuous updating of data on GS.

Individuals, institutions and countries are studied for their research output and impact. Choosing one of the 29 states in India, Garg and Kumar (2016) profiled the scientific output of scholars. Their study is about the research productivity of scientific institutions in a sprawling region in India. It covered all scientific publications produced by the scholars in the state of Odisha, and shows the possibilities of applying the same tools of scientometrics to HSS publications in similar contexts.

Scientometric studies forecast research trends and guide future research in disciplines and subjects. Drawing on publications ($n = 5,429$) from the *SSCI*, Tsai (2015) projected the future research trends and applications in e-commerce that have been developed over 20 years. Selected publications were classified into eight categories to check how e-commerce research trends and applications have been developed. The study tested the reliability of Lotka's law, a foundational law in scientometrics. The mapping showed that e-commerce is one of the fast-growing research topics, and countries such as the USA, China, Taiwan, England, South Korea and Canada have recorded growth in the area. Relevant disciplines for e-commerce subject categories were compiled, with computer science, business economics, engineering, information and library science, and operations research and management science being the main ones. This mapping also led to a list of journals that published an array of topics in e-commerce.

Here is another example that drew on the publications sourced directly from a journal. In full-text publications from a chosen journal in the discipline of human resource management, Ruggunan and Sooryamoorthy (2016) supplemented publication data with information on the race and gender of the authors. These extra variables were externally sourced from the personal and institutional websites of the authors. They were then combined with scientometric variables obtained from the publications, offering a new dimension to their study of their own and related disciplines (Case 4.16).

CASE 4.16 RACE, GENDER AND COLLABORATION IN HUMAN RESOURCE MANAGEMENT RESEARCH

Title of the Publication: Human resource management research in South Africa: A bibliometric study of authors and their collaboration patterns

Title of the Journal: *Journal of Contemporary Management*

Authors: Ruggunan, S. & Sooryamoorthy, R.

Institutional Affiliation: Human Resources Management; and Sociology, University of KwaZulu-Natal, South Africa

Year of Publication: 2016

Data Source: Publications in *South African Journal of Human Resources Management* for the period 2003–2013.

Research Objective/Questions: Examine the significance of race, gender, sector and institution in authorship and collaboration. Explore the demographic and institutional characteristics of authors who published in the selected journal.

Data, Methods and Analysis: All the articles (*n*=259) published in the journal during the period of analysis were accessed and downloaded. From these papers, variables such as the year of publication, institutional affiliation, collaboration and subject keywords were separated. The race and gender of authors were obtained from author/institutional websites and other internet sites. Racial and gender differences were found in the authorship of publications. Collaboration patterns of authors were strongly related to race, gender and institutional affiliation. Advanced statistical tests were applied. The characteristics of authors in the publications of the discipline of human resource management have also been elaborated.

Since information on gender or ethnicity is not readily available in the metadata of publications stored in citation indexes, scientometricians make efforts to find it from other sources. Scholars have devised methods to gather the information on the gender of authors. Some software like the gender guesser Python package (https://pypi.org/project/gender-guesser/) is effective. For a large-scale study involving thousands of records, the package, as employed by Draux et al. (2019), meets the objective. Science-Metrix (2018) developed a new approach to measure the proportion of women's authorship in scientific publications. The approach uses the given name and surname of authors to determine the possibility of the author being a man or woman. Zhang et al. (2019) obtained demographic information such as age and gender about Norwegian authors. While investigating gender disparities in economics for a large-scale study, Junwan et al. (2019) devised a mechanism to assign the gender of authors by using first names, following the method adopted by Larivière et al. (2013), who used a list of sources including the US Social Security database. Maddi et al. (2019) added the key variable of gender to their analysis of the disciplines of economics and management. They looked at the place of women in economics and management to see whether the difference in the levels of collaboration leads to joint publications. For this purpose, data was collected from 300 WoS-indexed journals in economics and 330 journals from management (79,078 and 90,222 articles, respectively). With the application of advanced statistical measures, the study confirmed that the practices of collaboration between men and women in the two disciplines are characteristically different. The additional data supplemented in this way meets the objective for higher levels of analysis.

In an extended scientometric study which appeared in book form, Ruggunan and Sooryamoorthy (2019) mapped the discipline of human resources management in South Africa. In this book, data from WoSCC and a human resources management journal was used. Full-text publications were downloaded from the website of the selected journal. The dataset was enriched with supplementary information on gender and race of the authors, which was collected from various internet sources. SPSS was used to capture, process and analyse the data. The attempt to trace the emergence and development of the discipline resulted in compiling its history over a period of time and in summarising its research foci. The characteristic features of the publications in this field, specific to a country context, were made known. A detailed sketch of the study is given in Case 4.17.

CASE 4.17 THE ORIGIN AND DEVELOPMENT OF MANAGEMENT STUDIES IN SOUTH AFRICA

Title of Book: *Management Studies in South Africa: Exploring the Trajectory in the Apartheid Era and Beyond*

Authors: Ruggunan, S. & Sooryamoorthy, R.

Institutional Affiliation: Human Resources Management and Sociology, University of KwaZulu-Natal, South Africa

Year of Publication: 2019

Data Source: WoSCC and publications from *South African Journal of Human Resources Management.*

Research Objective/Questions: Present the history and the patterns of knowledge production in the discipline of management studies in South Africa. Trace the development of the discipline into a profession and a practice and who (the background of authors in terms of race, gender and institutional affiliation) produces knowledge in this discipline.

Data, Methods and Analysis: The data was sourced from WoSCC and from a prominent journal in management studies. Bibliographic records of the publications of South African authors in the discipline of management were downloaded from WoSCC and entered into a data management program. Altogether, 1,294 publications covering the period of 1966–2015 were analysed. This WoSCC data was supplemented with data from the selected journal (*n*=259, published between 2005 and 2015). Full-text publications from the journal were used to extract relevant variables for analysis. In addition to these variables, the racial and gender backgrounds of the authors were collected separately from other sources. This information was combined to create the dataset. The analysis showed the unique features of the publications in management studies in relation to the race, gender, sector, institution, department and province of the authors. The relationship between these variables was revealed by applying inferential statistical tests. Both the historical and scientometric data were put together to present the history and development of the discipline in the country. What the study covered was the history of the discipline from the period of apartheid to the era of democracy.

In a similar approach, Sooryamoorthy (2016) adopted scientometric methods to construct the history of sociology in South Africa (Case 4.18). The research, from the colonial era to apartheid to democracy, captured the century-old history of the discipline in South Africa. Scientometric data collected from a group of journals and a citation index was combined with data on the gender and race of authors. Supplementary information on gender and race is crucial for a study like this in South Africa, which has a history of racial segregation. The strength of the work lies in the use of original scientometric data. As a reviewer wrote, it is "a thoroughly modern quantitative deconstruction of the history of a discipline" in a country (Waters, 2017). The strength of the book is also evident in the cross-tabulation of quantitative and qualitative data combined from journals and WoSCC (Katito, 2018). This is a study of a discipline by one who is part of it.

CASE 4.18 THE HISTORY AND DEVELOPMENT OF SOCIOLOGY IN SOUTH AFRICA

Title of Book: *Sociology in South Africa: Colonial, Apartheid and Democratic Forms*

Author: Sooryamoorthy, R.

Institutional Affiliation: Sociology Programme, School of Social Sciences, University of KwaZulu-Natal, South Africa

Year of Publication: 2016

Data Source: WoSCC and journals such as *Humanitas, Social Dynamics, South African Journal of Sociology, South African Sociological Review, South African Review of Sociology* (previously *Society in Transition*), *Transformation: Critical Perspectives on Southern Africa* and *Development Southern Africa.*

Research Objective/Questions: Trace the history and development of the discipline of sociology in South Africa under various political phases (colonialism, apartheid and democracy). Study the changing focus of the discipline in terms of research and teaching since 1900.

Data, Methods and Analysis: Scientometric data collected from WoSCC and selected sociology journals was combined with historical data sourced from reports, government gazettes, proceedings and other relevant historical documents. The lack of adequate coverage of local publications in the WoSCC database is balanced by the additional source of information. Additional and supplementary data was sourced from publications in the national journals in which South African scholars published their research. Moreover, data regarding the race and gender of the authors who published papers in national South African journals was collected and incorporated into the analysis. Collaboration of South African sociologists, both national and international, has also been investigated in the study. The methods of analysis were historical and quantitative. Scientometric data was statistically analysed and integrated into the historical analysis pertaining to each political period of colonialism, apartheid and democracy. Research areas of sociology for the different periods in the country have been extracted from scientometric data. Advanced statistical tests were employed. The characteristic features of sociology in South Africa, its developmental journey and the changing foci of sociological research and methodological preferences were revealed.

Some analyses, like the one shown in Case 4.19, overlap the boundaries of research evaluation and mapping. Nwagwu and Egbon, 2011) identified the characteristics of publications in Nigeria in the arts and the HSS obtained from WoS. It is a research evaluation study. The impact of the publications has also been assessed. It can be taken as a mapping study describing research in Nigeria over a period of time, presenting its international status of research in the arts and HSS disciplines. The authors come from the fields of information science and biological research.

CASE 4.19 PUBLICATIONS IN ARTS, HUMANITIES AND SOCIAL SCIENCES IN NIGERIA

Title of the Publication: Bibliometric analysis of Nigeria's social science and arts and humanities publications in Thomson Scientific databases

Title of the Journal: *The Electronic Library*

Authors: Nwagwu, W. & Egbon, O

Institutional Affiliation: Centre for Information Science, University of Ibadan, Nigeria; and the Nigerian Institute for Oil Palm Research, Nigeria

Year of Publication: 2011

Data Source: *A&HCI* and *SSCI*

Research Objective/Questions: Explore the extent of research conducted in the arts and HSS in Nigeria, by examining the outlets in which Nigerian scholars publish their research, and the international status of research publications.

Data, Methods and Analysis: The data, sourced from the citation indexes in WoS for the period 2002–2007, was analysed in terms of the number and type of publications; the sources/channels of publications; journal publishers and their location; authorship; and page, volume and number of references to the arts and HSS publications in Nigeria. Altogether, 716 publications were gathered. This descriptive study reveals the characteristics of the selected publications according to the above variables. Nigeria published very little research in the arts and HSS, the authors were not collaborative and the journals in which the publications appeared did not originate in Nigeria, but in England, the USA and Canada.

A similar work, but about Indian publications, has been produced. Tripathi and Babbar (2018) worked on the production of publications in the HSS disciplines in India, relying on data obtained from the *SSCI* and *A&HCI* for the past ten years. Pertaining to 9,525 records, the objectives were to explain the patterns of authorship, relationship between the number of authors and citations, the core journals in which the publications appeared, the main research areas of publications and the languages of these publications. Being an example of mapping of several disciplinary areas in the HSS, the strength of it is apparent in its coverage of a country. The scattering law of Bradford is tested with the data. Through this, comparable characteristics of publications produced in India and in the HSS disciplines were brought to light. There were eight core journals representing different disciplines in the HSS in which Indian scholars published most of their research. Notably, a majority of these journals originated in India. The relationship between the number of times the publications were accessed and the number of times they were cited has been probed. This is unique information on the publications that are accessed, read, used and cited. All the accessed or downloaded publications, as the study suggests, do not ordinarily result in citations. Nearly half of the publications have never earned any citations. References in the publications were also subjected to the analysis, which showed disciplinary variations and the importance of references and the count of references in research publications. In the HSS, it is expected that publications belong to more than one disciplinary area. The extent of multidisciplinarity in the HSS, as has been explored, is valuable to any mapping study concerning these disciplines. Inferential statistical procedures were applied to search for the relationship between the chosen variables and to test hypotheses.

Taking data from the *SSCI*, Köseoglu et al. (2015) analysed publication records to describe the scholarship in tourism and hospitality and its development in Turkey during a 30-year period (1984–2013). Köseoglu et al. pursued the trends and patterns in the discipline, which is a common objective of any mapping research. The methodology adopted was meant to collect the publications in the field with the title of the journal, year of publication, author names, authorship information, institutional contribution, themes of research, discipline of the topic and research methods. Also gathered was information on the references used in the publications, which was intended to find the characteristics of citations, influential scholars, highly cited publications and other important publications. This research has led to some typical features of the discipline. The overall representation of Turkey in the international literature on tourism and hospitality is limited. Many of the study's findings can lead to an in-depth understanding of the current status of the discipline and can suggest ways towards its future development.

Again from Turkey, Al et al. (2006) examined the publications of scholars in the humanities fields affiliated to Turkish institutions. The focus is on the journal publications in the arts and humanities subjects to illustrate the distribution of publications (types), journals in which the papers were published (core journals) and referencing patterns. The *A&HCI* covering the period of 1975–2003 was resorted to. The distinguishing features of these publications in terms of the language (English or other), single/coauthorship, references in the articles (papers/monographs), age of references and citation behaviour were found. The increase/decrease in the number of publications, the ratio of Turkish publications to world publications in the selected fields, the core journals in which papers were published, the proportion of single- and coauthored papers, and the typical characteristics of referencing were made known.

Another scientometric survey of social sciences literature in Turkey was conducted. Uzun (1998) portrays the research activities relating to the social sciences in Turkey. In line with the tracking of disciplines, the survey revealed the growth of publication outputs, distribution of publications in journals, authorship patterns and subjects.

In a unique way, book reviews have been taken as a source material to map disciplinary trends. Acknowledging scholarly book reviews as significant channels of scholarly communication, Lindholm-Romantschuk (1998) examined them to map the disciplinary knowledge flows in the HSS. Zuccala et al. (2019) employed book reviews from WoSCC to show the top book disciplines that received book reviews of ten or more per book, and their subject area flow.

From the above collection of cases and examples, the mapping of disciplines in a country, region or institution is a fruitful exercise in scientometrics. As part of mapping, a range of aspects, from production output to the structure of a discipline to the emergence of a new field of study, is examined by scholars at the micro and macro levels. A variety of data sources is used. The methodologies adopted also present a range of possibilities for future studies. The variables identified and used in these cases and examples are not limited to a few. Some have integrated additional information on gender, race and age, which are not included in citation indexes or in publications. Statistical combinations of these variables, as described in some of the cases and examples discussed above, expose the complexities of disciplines in terms of origin, history, trends, growth, stagnation, decline and evolution. Some cases and examples in this section show how a new emerging field or subject of study and the forecasting of research trends can be outlined.

Citation analyses

In comparison to the science disciplines, there are significant variations in the use of resources for the HSS disciplines. Investigating citation patterns in some of the major fields/disciplines in the humanities, Knievel and Kellsey (2005) reported on the characteristic patterns that vary from field to field. Librarians and bibliographers by profession, the authors counted the references in the papers published in a major journal in each of the fields. References, articles and pages, references per article and page, and monographs in English and in foreign languages were analysed for each of the fields. The findings, as summarised in Case 4.20, refer not only to the humanities in general, but also to the fields (disciplines) within the humanities and show how they differ in the use of references (books versus journals and the language of the resources). The practical application of this study is mainly for libraries to make decisions on the purchase of books and journals for each of these fields. Some fields require more book publications while others continue to maintain their journal subscriptions. Studies of this kind can advise decision makers on the allocation of library resources to higher learning institutions.

CASE 4.20 CITATION PATTERNS IN THE HUMANITIES

Title of the Publication: Citation analysis for collection development: A comparative study of eight humanities fields

Title of the Journal: *The Library Quarterly: Information, Community, Policy*

Authors: Knievel, J. E. & Kellsey, C.

Institutional Affiliation: Humanities Reference and Instruction Librarian, and Monographic Cataloguer and Bibliographer, University of Colorado at Boulder, USA

Year of Publication: 2005

Data Source: One major journal representing each of the selected fields in the humanities published in 2002.

Research Objective/Questions: Analyse the citation patterns among the chosen eight humanities fields (arts, classics, history, linguistics, literature, music, philosophy and religion). Investigate the use of foreign-language resources by scholars and the relative percentages of books and journals cited.

Data, Methods and Analysis: The citations of all selected journals for the year 2002 were counted separately for each of the fields. Manually computed, the citations (n=9,131) were then entered into spreadsheets. The citation patterns among the selected fields in the humanities were found to be varied. About three quarters of the citations were books and more than three quarters of them were English-language sources. The highest average reliance on books (more than 80%) was obtained in the fields of literature, music and religion. When the language of the sources cited in the fields was considered, foreign-language citations were dominated by both French and German. No single foreign language was dominant in all humanities fields. The number of citations per article differed from field to field, the lowest being reported in philosophy and the highest in arts.

In a citation analysis of social science journals in Taiwan, Tsay (2015) pursued citations across the type of document, language and year of publication. The index used was the *Taiwan Social Sciences Citation Index*. Several of the findings point to the differing features of social science literature. The analysis was segregated by disciplines such as sociology, education, psychology, political science, economics and management. Citation types in these journals were grouped according to location and type in the publications. The locations included the setting of the study, background information and methodological sections, while the types of citation were argumentative, hypothetical, historical or casual. Tsay found from the analysed journals that journals and books are the most cited, English is the predominant language of citations, the age of citations is around ten years and the citations are related to the citation location in the publication.

The significance of books and monographs in the research communication of the HSS has led to their increased uses. Books and monographs are not adequately covered in major databases. Some works that compared citations between books and journals are available in the literature. By adopting Google Books search (GBS), Kousha and Thelwall (2009) explored citation patterns in ten science and HSS disciplines. GBS has turned out to be a valuable resource of citations in the HSS. According to the study, in some social science disciplines books play a significant role in scholarly communication (Case 4.21).

CASE 4.21 CITATIONS IN THE HSS USING GOOGLE BOOKS SEARCH

Title of the Publication: Google book search: Citation analysis for social science and the humanities

Title of the Journal: *Journal of the American Society for Information Science and Technology*

Authors: Kousha, K. & Thelwall, M.

Institutional Affiliation: Department of Library and Information Science, University of Tehran, Iran; and the School of Computing and Information Technology, University of Wolverhampton, UK

Year of Publication: 2009

Data Source: GBS and WoSCC

Research Objective/Questions: Examine whether GBS is a useful resource for the citation impact assessment of academic journal articles; whether GBS citations of journal articles correlate with their WoSCC citations at the article and journal levels; and whether disciplinary differences influence citations in the HSS.

Data, Methods and Analysis: All the research articles ($n = 1,923$) published in the 51 information science and library science journals indexed by WoS were analysed. The number, mean and median of GBS citations were compared with WoSCC article and journal citations. A random sample of articles from WoS-indexed journals to represent science and the HSS was taken to compare disciplinary similarities and dissimilarities. Apart from descriptive statistics, correlation tests were performed to analyse the data. In the HSS disciplines, the number, mean and median of the GBS citations were much higher than those of the science disciplines. This is an indication that GBS had a good coverage of books for citation analysis in the HSS, and that GBS citations were valuable for the impact assessment of research articles in the HSS. A strong relationship between GBS citations and WoSCC citations in all selected disciplines was observed. However, clear differences between the GBS and WoSCC citations were evident. For disciplines in which books play a vital role in communication, GBS can be used as an appropriate source of citation data for research evaluation.

Not only citations of disciplines or subject fields, but also the topics within a discipline can be separated for citation analysis. Hladík's (2019) unique attempt to use automated topic modelling was found in the examination of the cumulative citation counts of topical words in sociology. The search is to see whether the topical structure of a discipline can have an effect on citation counts. With full-text data obtained from the journal *Czech Sociological Review*, and citation data from WoSCC, the effect of the sociological topics on citations was assessed. The models of Hladík indicated that quantitative topics and social geography yield a more considerable advantage in citations than others. The opposite effects for theoretical and quantitative topics in sociology showed a disconnect between theory and empirical research in Czech sociology. This is a case in which sophisticated application of scientometric methods was adopted.

Studies demand the need for the investigation of multiple references citations and unitary reference citations in publications. A multiple references citation is a citation that includes more than one reference of the same citation in a publication, whereas a unitary reference citation is a citation that is used only once in a publication. Lin et al. (2019) examined both citation types, their location in the full text, the shares of self-citing citation and their citation age. The method contributes to the measurement of the percentage of multiple references and unitary reference citations, their locations in the text (beginning, middle or end), the age of references and their correlations. Both discipline-specific and author-specific features of citations can be deduced from this type of scientometric analysis.

From the above cases and examples of citation analysis, one can see which aspects of citations are important in scientometric studies. Some have delved deeply into the references cited in the publications, measuring the types of references (journals, chapters and monographs), number of references, the page length of articles, the age of the references and the language of references cited across disciplines in the HSS.

Collaboration and coauthorship

Collaboration is a well-researched area of investigation in scientometrics, but it is not as common in the HSS as in science in which it is a normal practice. Scientometric analysis of collaboration among sociologists by sociologists is relevant in two respects in the study carried out by Hunter and Leahey (2008): it uses publications from two key journals in sociology; and additional information on gender and institutional prestige has been incorporated into the analysis (Case 4.22). By examining collaboration and its various contributing factors, this overlaps with a mapping study. Being a key variable in scientometrics, collaboration has more to offer in disclosing the complex dynamics of research in a discipline. Collaboration studies like this are effectively done with publications collected directly from journals and not necessarily from citation indexes.

CASE 4.22 FINDING COLLABORATION AMONG SOCIOLOGISTS USING SPECIFIC JOURNALS

Title of the Publication: Collaborative research in sociology: Trends and contributing factors

Title of the Journal: *The American Sociologist*

Authors: Hunter, L. & Leahey, E.

Institutional Affiliation: Department of Sociology, University of Arizona, USA

Year of Publication: 2008

Data Source: *American Journal of Sociology* and *American Sociological Review*

Research Objective/Questions: Explore the extent, change and factors of collaboration among sociologists over a 70-year period.

Data, Methods and Analysis: Data from the above two journals ($n = 1,274$ publications) for the period of 1935–2005 were collected through stratified (by issue of the journals) random sampling (20%). A dependent variable (collaboration) and explanatory variables (year of publication, gender, institutional affiliation, department prestige, location, methodological approach in publications and source of primary/secondary data) were used. Additional information about the gender of authors and institutional prestige was gathered separately and integrated with the data already available. Advanced statistical methods including regression were employed. Results are presented diagrammatically as well. According to the study, the importance of geographical location in collaboration had been declining since 1950; cross-place collaborations stagnated during the later years; quantitative research attracted more collaboration; there were no gender differences in collaboration; institutional prestige was higher in collaborated publications than in sole-authored publications; and sole authorship remained the common form of authorship.

Mena-Chalco et al. (2014) investigated coauthorship patterns among Brazilian authors in fields that included the HSS. Unlike the conventional use of databases, in their study Mena-Chalco et al. made use of the Brazilian Lattes Platform to see the networks of coauthorships. The platform carries data on the academic activities of researchers associated with knowledge areas. Not represented sufficiently in prominent citation databases, Brazilian publications amount to a significant part of world knowledge. The structure and dynamics of collaboration among researchers in all major areas, sciences and the HSS, are exposed through this analysis. Agricultural sciences, biological sciences, earth sciences, humanities, applied social sciences, health sciences, engineering, linguists and arts exhibited differing behaviours in coauthorship relations.

By investigating coauthorship patterns in the social sciences for about three decades, Henriksen (2015) shows how different research areas and methodological differences lead to coauthorship tendencies. A large number of articles published in 56 research areas in WoS were taken as the sample. In the evolution of coauthorships in the social sciences, a substantial increase was found in the majority of the research areas. This increasing tendency of coauthorship has implications for collaboration, internationalisation and citations. Focusing on the anthropology, economics, history, political science, psychology and sociology disciplines, Sangam and Keshava (2005) revealed the coauthorship patterns among Indian scholars. Except for the political science and sociology disciplines, a consistent growth in coauthorships was apparent in the publications.

By mining data from the Flemish Academic Bibliographic Database for the Social Sciences and Humanities, Ossenblok et al. (2014) pursued coauthorship patterns in the HSS in the recent decade. Journal publications and book chapters in 16 HSS disciplines were included in the database. In the evolution of coauthorship, as manifested in the publications of the HSS in Flanders, there has been an increasing number of collaborative publications. A comparison with the publications indexed in WoS was also made, which showed that WoS had a higher number of coauthors than in the journals not indexed by WoS and book chapters. Between the humanities and the social sciences, collaboration was more prominent in the former than in the latter.

Despite these characteristic publication practices of the HSS, monographs and books do not appear as subjects of scientometric studies as do journal publications. In this rare group of studies, Ossenblok and Engels (2015) looked at the peer-reviewed edited books published in the 16 disciplines in the HSS in Flanders, the Dutch-speaking part of Belgium. While searching for the features of edited books, the authors were interested in exploring the collaboration patterns of scholars (Case 4.23). The characteristics of these publications were plotted with information on the distribution of publishers, places of publication, language, the structure of chapters (presence of introductions and conclusions), coauthorship and the count of chapters.

CASE 4.23 COLLABORATION IN THE HSS THROUGH AN ANALYSIS OF BOOK PUBLICATIONS

Title of the Publication: Edited books in the social sciences and humanities: Characteristics and collaboration analysis

Title of the Journal: *Scientometrics*

Authors: Ossenblok, T. L. B. & Engels, T. C. E.

Institutional Affiliation: Centre for Research & Development Monitoring (ECOOM), Faculty of Political and Social Sciences, University of Antwerp, Belgium; and Antwerp Maritime Academy, Belgium

Year of Publication: 2015

Data Source: The Flemish Academic Bibliographic Database for the Social Sciences and Humanities (VABB-SHW), data harvested online and from the university libraries.

Research Objective/Questions: Analyse the characteristics and collaboration patterns of edited books in the HSS by studying the distribution of publishers, places of publication, language, coeditorship and coauthorships, and the presence of introductions and conclusions in the chapters.

Data, Methods and Analysis: The sample consisted of 753 peer-reviewed edited books and 12,913 chapters in 16 disciplines in the HSS. Bibliographic information was collected according to the chosen variables. Missing information for the variables was collected manually. The variables were then entered into a database. The revised collaborative coefficient, which takes into account the total number of papers and authors, and papers having a certain number of authors, was calculated. Collaboration was investigated in detail. Statistical tests such as the Kruskal-Wallis test, Chi-square test and correlation tests were employed to check the relationship between collaboration and disciplines. Apart from the general characteristics of edited books, the study found that about half of the books were published with about 5 per cent of the publishers; English was the dominant language of publication; 90 per cent of the edited books were coedited; and edited books in the social sciences had a more diverse authorship than the edited books in the humanities.

Gender is a key variable in scientometric analysis. Aksnes et al. (2019) conducted a large-scale study on gender differences in international research collaboration among Norwegian researchers. They covered both the HSS and the natural sciences (Case 4.24). Based on the analysis of coauthored publications, the focus was on the gender differences in international research collaboration. Indicators constructed for the study were the share of researchers involved in international collaboration, and the share of their publications with international coauthorship.

CASE 4.24 GENDER DIFFERENCES IN RESEARCH COLLABORATION

Title of the Publication: Gender gaps in international research collaboration: A bibliometric approach

---

Title of the Journal: *Scientometrics*

Authors: Aksnes, D. W., Piro, F. N. & Rørstad, K.

Institutional Affiliation: Nordic Institute for Studies in Innovation, Research and Education (NIFU), Norway

Year of Publication: 2019

Data Source: The CRIStin database (the Norwegian Scientific Index) which has been developed for all research institutions in Norway. The database was then coupled with the Norwegian Research Personnel Register, which holds personal information of individual scholars relating to gender, age, position and institution.

Research Objective/Questions: Find gender differences in international research collaboration by comparing international research coauthorship among men and women scholars at Norwegian universities.

Data, Methods and Analysis: The database used in the study has a full coverage of all peer-reviewed scientific and scholarly publications such as articles, books, edited volumes and conference series. Researchers (n = 5,554) from four major universities in Norway and their research output for the period 2015–2017 were analysed. Research collaboration was examined at the level of gender, fields, disciplines and academic positions. Findings indicated that women published less than men in all fields, but this underrepresentation of women was moderate in the HSS. Gender difference in international research collaboration was noticeable. International collaboration was much more frequent in the natural sciences, medical and health sciences, and technology than in the HSS. Association between the volume of publications and international collaboration was reported. Advanced statistical techniques were applied to the data.

Being the most searched area in scientometric studies, collaboration and coauthorship can be either on their own or part of a mapping or research evaluation study. Along with the information on the type of collaboration (local, national, regional or international), these studies also contribute to important areas of analysis. Gender differences in collaboration, internationalisation of disciplines and the extent of inter- and multidisciplinarity of disciplines in the HSS are notable among them.

The above collection of cases and examples of scientometric studies on a range of disciplines in the HSS points to the extensive use of scientometrics for a variety of purposes. The potential for a more creative use of scientometrics for purposes such as mapping, research evaluation, impact studies, citation analysis and authorships is reflected in these studies. The distinguishing publication practices in the HSS demand an approach which is not generally followed in science. The cases and examples explain both the extent of major data sources and unconventional datasets that can be purposefully adopted in the HSS. Not only major datasets, but also lesser known local citation indexes and journals are reliable repositories of necessary data. Gender, race and age – not normally featured in scientometric analysis – have been integrated into the analyses. These variables add values to the study of the complex heterogeneous nature and characteristics of disciplines in the HSS. The institutional and disciplinary affiliations of the authors provided in the cases signify who is interested in which disciplines in the HSS.

Many more areas remain to offer the avenues of investigation for scientometric research for social scientists. Included among them are:

- Qualitative data drawn from the records of publications (abstracts, keywords, titles of the papers to identify methodological preferences, subject areas, changing research focus, strong and weak disciplines, subjects, fields and contexts, countries, regions and world).
- Production of knowledge (publication trends) in various fields/disciplines within the HSS.
- Knowledge production in relation to collaboration (institutional, national, regional or international).
- Production of knowledge through the analysis of publication outlets and their origin (local/national/international).
- Networks of collaboration within disciplines and interdisciplinary, using information on departmental affiliation of authors.
- Authorship patterns (the order of author, the significance of the order for the dominance of authors) and the relationship with institutions, funding, country, disciplines, subjects and outlets of publications.
- Funding – who funds from where, north to south or cofunding.
- Citations – relationship with subject areas, collaboration, number of authors (whether more collaborations yield more citations or not).
- Growth and development of disciplines in the HSS.
- Specific areas of growth/decline in a particular discipline within the HSS.
- Relationship between departmental affiliations and subject areas of authors.
- Productivity of authors over the years to see whether their foci have changed.
- Authors and demographic variables (race, gender and age) that can be supplemented with data from other sources.
- Comparison between countries in the production of knowledge in specific disciplines of the social sciences.
- Country-specific analysis of the HSS disciplines, and which discipline grew/declined over the years.

Having seen the use, applications, methodologies and the potential of scientometric studies in the HSS, attention should now turn to the sources of data and its processing and analysis, which are detailed in the next chapter.

References

Aaltojärvi, I., Arminen, I., Auranen, O. & Pasanen, H.-M. (2008). Scientific productivity, web visibility and citation patterns in sixteen Nordic sociology departments. *Acta Sociologica*, *51*, 5–22. https://doi.org/10.1177%2F0001699307086815

Aksnes, D.W., Piro, F. N. & Rørstad, K. (2019). Gender gaps in international research collaboration: A bibliometric approach. *Scientometrics*, *120*, 747–774. https://doi.org/10.1007/s11192-019-03155-3

Al, U., Sahiner, M. & Tonta, Y. (2006). Arts and humanities literature: Bibliometric characteristics of contributions by Turkish authors. *Journal of the American Society for Information Science and Technology*, *57*, 1011–1022. https://doi.org/10.1002/asi.20366

Anglada-Tort, M. & Sanfilippo, K. R. M. (2019). Visualizing music psychology: A bibliometric analysis of psychology of music, music perception, and musicae scientiae from 1973 to 2017. *Music & Science*, *2*, DOI: 10.1177/2059204318811786.

Barrios, M., Borrego, A., Vilaginés, A., Ollé, C. & Somozab, M. (2008). A bibliometric study of psychological research on tourism. *Scientometrics*, *77*, 453–467. DOI: 10.2307/2799267

Benckendorff, P. & Zehrer, A. (2013). A network of analysis of tourism research. *Annals of Tourism Research*, *13*, 121–149. https://doi.org/10.1016/j.annals.2013.04.005

Biglu, M.-H., Chakhmachi, N. & Biglu, S. (2013). Scientific study of Middle East countries in psychology (1996–2010). *Collnet Journal of Scientometrics and Information Management*, *7*, 293–296. https://doi.org/10.1080/09737766.2013.832900

Bjarnason, T. & Sigfusdottir, I. D. (2002). Nordic impact: Article productivity and citation patterns in sixteen Nordic sociology departments. *Acta Sociologica*, *45*, 253–267. https://doi.org/10.1177/000169930204500401

Bornmann, L., Thor, A., Marx, W. & Schier, H. (2016). The application of bibliometrics to research evaluation in the humanities and social sciences: An exploratory study using normalized Google Scholar data for the publications of a research institute. *Journal of the American Society for Information Science and Technology*, *67*, 2778–2789. https://doi.org/10.1002/asi.23627

Caillods, F. (2016). Knowledge divides: Social science production on inequalities and social justice. In ISSC, IDS & UNESCO, *World Social Science Report 2016. Challenging Inequalities: Pathways to a Just World* (pp. 280–285). Paris: UNESCO Publishing. https://unesdoc.unesco.org/ark:/48223/pf0000245891

Chandra, Y. (2017). Mapping the evolution of entrepreneurship as a field of research (1990±2013): A scientometric analysis. *PLOS One*, *13*, https://doi.org/10.1371/journal.pone.0190228.

Chi, P.-S. (2012). Bibliometric characteristics of political science research in Germany. *Proceedings of the American Society for Information Science and Technology*, *49*, 1–6. https://doi.org/10.1002/meet.14504901115

Diem, A. & Wolter, S. C. (2013). The use of bibliometrics to measure research performance in education sciences. *Research in Higher Education*, *54*, 86–114. https://doi.org/10.1007/s11162-012-9264-5

Draux, H., Porter, S., Beck, R., Kundu, S. & Konkiel, S. (2019). Visualizing gender representation by field of research at institutions in the United Kingdom. In G. Catalano, C. Daraio, M. Gregori, H. F. Moed & G. Ruocco (Eds), *Proceedings of the 17th Conference of the International Society for Scientometrics and Informetrics* (Vol. II, pp. 2503–2504). Rome: Edizioni Efesto.

Erfanmanesh, M., Zohoorian-Fooladi, N. & Abrizah, A. (2019). The ASEAN University network research performance: A meso-level scientometric assessment. In G. Catalano, C. Daraio, M. Gregori, H. F. Moed & G. Ruocco (Eds), *Proceedings of the 17th Conference of the*

International Society for Scientometrics and Informetrics (Vol. I, pp. 99–111). Rome: Edizioni Efesto.

Garg, K. C. & Kumar, S. (2016). Scientometric profile of an Indian state: The case study of Odisha. *Collnet Journal of Scientometrics and Information Management*, *10*, 141–153. https://doi.org/10.1080/09737766.2016.1177950

Gingras, Y. & Mosbah-Natanson, S. (2010). Where are social sciences produced? In UNESCO & ISSC (Eds), *World Social Science Report 2010: Knowledge Divides* (pp. 149–153). Paris: UNESCO & ISSC.

Godin, B. (2002). *The Social Sciences in Canada: What Can We Learn from Bibliometrics. Working Paper No. 1.* Quebec: INRS, University of Quebec. www.csiic.ca/PDF/CSIIC.pdf

Gumpenberger, C., Sorz, J., Wieland, M. & Gorraiz, J. (2016). Humanities and social sciences in the bibliometric spotlight – Research output analysis at the University of Vienna and considerations for increasing visibility. *Research Evaluation*, *25*, 271–278. https://doi.org/10.1093/reseval/rvw013

Henriksen, D. (2015). The rise in co-authorship in the social sciences (1980–2013). In A. A. Salah, Y. T. A. A. A. Salah, C. Sugimoto & U. Al (Eds), *Proceedings of ISSI 2015 Istanbul: 15th International Society of Scientometrics and Informetrics Conference* (pp. 209–220). Istanbul: Boğaziçi University Printhouse.

Hladík, R. (2019). Why sociologists should not bother with theory: The effect of topics on citations. In G. Catalano, C. Daraio, M. Gregori, H. F. Moed & G. Ruocco (Eds), *Proceedings of the 17th Conference of the International Society for Scientometrics and Informetrics* (Vol. II, pp. 2341–2346). Rome: Edizioni Efesto.

Ho, H.-C. & Ho, Y.-S. (2015). Publications in dance field in Arts & Humanities Citation Index: A bibliometric analysis. *Scientometrics*, *105*, 1031–1040. https://doi.org/10.1007/s11192-015-1716-1

Huang, M.-h. & Chang, Y.-w. (2008). Characteristics of research output in social sciences and humanities: From a research evaluation perspective. *Journal of the American Society for Information Science and Technology*, *59*, 1819–1828. https://doi.org/10.1002/asi.20885

Hunter, L. & Leahey, E. (2008). Collaborative research in sociology: Trends and contributing factors. *The American Sociologist*, *39*, 290–306. https://doi.org/10.1007/s12108-008-9042-1

Junwan, L., Yinglu, S., Sai, Y., Sugimoto, C. r. & Lariviere, V. (2019). Gender disparities in the field of economics. In G. Catalano, C. Daraio, M. Gregori, H. F. Moed & G. Ruocco (Eds), *Proceedings of the 17th Conference of the International Society for Scientometrics and Informetrics* (Vol. I, pp. 932–943). Rome: Edizioni Efesto.

Katito, J. (2018). Sociology in South Africa. Book review. *International Sociology Reviews*, *33*, 611–614.

Knievel, J. E. & Kellsey, C. (2005). Citation analysis for collection development: A comparative study of eight humanities fields. *The Library Quarterly: Information, Community, Policy*, *75*, 142–168. DOI: 10.1086/431331

Köseoglu, M. A., Rahimi, R., Okumus, F. & Liu, J. (2016). Bibliometric studies in tourism. *Annals of Tourism Research*, *61*, 180–198. https://doi.org/10.1016/j.annals.2016.10.006

Köseoglu, M. A., Sehitogluc, Y. & Parnell, J. A. (2015). A bibliometric analysis of scholarly work in leading tourism and hospitality journals: The case of Turkey. *Anatolia – An International Journal of Tourism and Hospitality Research*, *26*, 359–371. https://doi.org/10.1080/13032917.2014.963631

Kousha, K. & Thelwall, M. (2009). Google book search: Citation analysis for social science and the humanities. *Journal of the American Society for Information Science and Technology*, *60*, 1537–1549. https://doi.org/10.1002/asi.21085

Lancho-Barrantes, B. S. & Cantu-Ortiz, F. J. (2019). Quantifying the research preferences of top research universities: Why they make a difference? In G. Catalano, C. Daraio, M. Gregori, H. F. Moed & G. Ruocco (Eds), *Proceedings of the 17th Conference of the International Society for Scientometrics and Informetrics* (Vol. II, pp. 1488–1499). Rome: Edizioni Efesto.

Larivière, V., Ni, C., Gingras, Y., Cronin, B. & Sugimoto, C. R. (2013). Bibliometrics: Global gender disparities in science. *Nature, 504,* 211–213. DOI: 10.1038/504211a

Larsen, P. O., Maye, I. & Ins, M. v. (2008). Scientific output and impact: Relative positions of China, Europe, India, Japan and the USA. *Collnet Journal of Scientometrics and Information Management, 2,* 1–10. DOI: 10.1080/09737766.2008.10700848

Lewison, G., Abouzeid, M., Sabouni, A., Zalabany, M. E., Jabbour, S. & Sullivan, R. (2019). Identification of the research on warfare and health, 1946–2017. In G. Catalano, C. Daraio, M. Gregori, H. F. Moed & G. Ruocco (Eds), *Proceedings of the 17th Conference of the International Society for Scientometrics and Informetrics* (Vol. I, pp. 271–282). Rome: Edizioni Efesto.

Lin, G., Hou, H. & Hu, Z. (2019). Understanding multiple references citation. In G. Catalano, C. Daraio, M. Gregori, H. F. Moed & G. Ruocco (Eds), *Proceedings of the 17th Conference of the International Society for Scientometrics and Informetrics* (Vol. I, pp. 2347–2357). Rome: Edizioni Efesto.

Lindholm-Romantschuk, Y. (1998). *Scholarly Book Reviewing in the Social Sciences and Humanities: The Flow of Ideas Within and Among Disciplines.* Westport, CT: Greenwood Press.

Liu, X., Hong, S. & Liu, Y. (2012). A bibliometric analysis of 20 years of globalization research: 1990–2009. *Globalizations, 9,* 195–210. https://doi.org/10.1080/14747731.2012.658256

Maddi, A., Larivière, V. & Gingras, Y. (2019). Man-woman collaboration practices and scientific visibility: How gender affects scientific impact in economics and management. In G. Catalano, C. Daraio, M. Gregori, H. F. Moed & G. Ruocco (Eds), *Proceedings of the 17th Conference of the International Society for Scientometrics and Informetrics* (Vol. I, pp. 1687–1697). Rome: Edizioni Efesto.

Mena-Chalco, J. P., Digiampietri, L. A., Lopes, F. M. & Junior, R. M. C. (2014). Brazilian bibliometric coauthorship networks. *Journal of the American Society for Information Science and Technology, 65,* 1424–1445. DOI: 10.1002/asi.23010

Mosbah-Natanson, S. & Gingras, Y. (2014). The globalization of social sciences? Evidence from a quantitative analysis of 30 years of production, collaboration and citations in the social sciences (1980–2009). *Current Sociology, 62,* 626–646. https://doi.org/10.1177/0011392113498866

Mutz, R. & Daniel, H.-D. (2019). How should we measure individual researchers' performance capacity within and between universities – Social sciences as an example? A multilevel extension of the bibliometric quotient (BQ). In G. Catalano, C. Daraio, M. Gregori, H. F. Moed & G. Ruocco (Eds), *Proceedings of the 17th Conference of the International Society for Scientometrics and Informetrics* (Vol. I, pp. 1098–1109). Rome: Edizioni Efesto.

Nwagwu, W. & Egbon, O. (2011). Bibliometric analysis of Nigeria's social science and arts and humanities publications in Thomson Scientific databases. *The Electronic Library, 29,* 438–456. DOI: 10.1108/02640471111156722

Ossenblok, T. L. B. & Engels, T. C. E. (2015). Edited books in the social sciences and humanities: Characteristics and collaboration analysis. *Scientometrics, 104,* 219–237. https://doi.org/10.1007/s11192-015-1544-3

Ossenblok, T. L. B., Verleysen, F. T. & Engels, T. C. E. (2014). Coauthorship of journal articles and book chapters in the social sciences and humanities (2000–2010). *Journal of the American Society for Information Science and Technology, 65,* 882–897. https://doi.org/10.1002/asi.23015

Perron, B. E., Victor, B. G., Hodge, D. R., Salas-Wright, C. P., Vaughn, M. G. & Taylor, R. J. (2017). Laying the foundations for scientometric research: A data science approach. *Research on Social Work Practice*, *27*, 802–812. https://doi.org/10.1177/1049731515624966

Phelan, T. J. (2000). Bibliometrics and the evaluation of Australian sociology. *Journal of Sociology*, *36*, 345–363. DOI: 10.1177/144078330003600305

Pölönen, J. & Hammarfelt, B. (2019). Historical bibliometrics using Google Scholar: The case of Roman law, 1500–2016. In G. Catalano, C. Daraio, M. Gregori, H. F. Moed & G. Ruocco (Eds), *Proceedings of the 17th Conference of the International Society for Scientometrics and Informetrics* (Vol. I, pp. 2491–2492). Rome: Edizioni Efesto.

Prins, A. A. M., Costas, R., van Leeuwen, T. N. & Wouters, P. F. (2016). Using Google Scholar in research evaluation of humanities and social science programs: A comparison with Web of Science data. *Research Evaluation*, *25*, 264–270. https://doi.org/10.1093/reseval/rvv049

Ruggunan, S. & Sooryamoorthy, R. (2016). Human resource management research in South Africa: A bibliometric study of authors and their collaboration patterns. *Journal of Contemporary Management*, *13*, 1394–1427.

Ruggunan, S. & Sooryamoorthy, R. (2019). *Management Studies in South Africa: Exploring the Trajectory in the Apartheid Era and Beyond*. Basel, Switzerland: Springer.

Sandström, U. & Besselaar, P. v. d. (2019). Performance of research teams: Results from 107 European groups In G. Catalano, C. Daraio, M. Gregori, H. F. Moed & G. Ruocco (Eds), *Proceedings of the 17th Conference of the International Society for Scientometrics and Informetrics* (Vol. I, pp. 2240–2251). Rome: Edizioni Efesto.

Sangam, S. L. & Keshava. (2005). Collaboration in social science research in India. In P. Ingwersen & B. Larsen (Eds), *Proceedings of ISSI 2005– The 10th International Conference of the International Society for Scientometrics and Informetrics* (Vol. 1, pp. 775–778). Stockholm: Karolinska University Press.

Science-Metrix. (2018). *Analytical Support for Bibliometrics Indicators: Development of Bibliometric Indicators to Measure Women's Contribution to Scientific Publications*. Montréal: Science-Metrix Inc.

Serenko, A., Bontis, N. & Grant, J. (2009). A scientometric analysis of the proceedings of the McMaster World Congress on the Management of Intellectual Capital and Innovation for the 1996–2008 period. *Journal of Intellectual Capital*, *10*, 8–21. https://doi.org/10.1108/14691930910922860

Singh, V. K., Banshal, S. K., Singhal, K. & Uddin, A. (2015). Scientometric mapping of research on "Big Data". *Scientometrics*, *105*, 727–741. https://doi.org/10.1007/s11192-015-1729-9

Sooryamoorthy, R. (2016). *Sociology in South Africa: Colonial, Apartheid and Democratic Forms*. Hampshire and New York: Palgrave Macmillan.

Sorz, J., Glänzel, W., Ulrych, U. & Gorraiz, J. (2019). Institutional research specializations identified by esteem factors and bibliometric means: A case study at the University of Vienna. In G. Catalano, C. Daraio, M. Gregori, H. F. Moed & G. Ruocco (Eds), *Proceedings of the 17th Conference of the International Society for Scientometrics and Informetrics* (Vol. I, pp. 873–884). Rome: Edizioni Efesto.

Tripathi, M. & Babbar, S. K. P. (2018). Bibliometrics of social science and humanities research in India. *Current Science*, *114*, 2240–2247. DOI: 10.18520/cs/v114/i11/2240–2247

Tsai, H.-H. (2015). The research trends forecasted by bibliometric methodology: A case study in e-commerce from 1996 to July 2015. *Scientometrics*, *105*, 1079–1089. https://doi.org/10.1007/s11192-015-1719-y

Tsay, M.-y. (2015). Citation type analysis for social science literature in Taiwan. In A. A. Salah, Y. T. A. A. A. Salah, C. Sugimoto & U. Al (Eds), *Proceedings of ISSI 2015 Istanbul: 15th International Society of Scientometrics and Informetrics Conference* (pp. 117–128). Istanbul: Boğaziçi University Printhouse.

Uzun, A. (1998). A scientometric profile of social sciences research in Turkey. *The International Information & Library Review*, 30, 169–184. https://doi.org/10.1080/10572317.1998.10762473

van Leeuwen, T. (2013). Publication trends in social psychology journals: A long-term bibliometric analysis. *European Journal of Social Psychology*, *43*, 9–11. https://doi.org/10.1002/ejsp.1933

Verleysen, F. T. & Engels, T. C. E. (2014). Internationalization of peer reviewed and non-peer reviewed book publications in the social sciences and humanities. *Scientometrics*, *101*, 1431–1444. https://doi.org/10.1007/s11192-014-1267-x

Waters, G. (2017). Sociology in South Africa. Book review. *Transformation: Critical Perspectives on Southern Africa*, *93*, 170–175.

Zhang, L., Du, H., Y., H., Glänzel, W. & Sivertsen, G. (2019). Gender, age, and broader impact: A study of persons, not just authors. In G. Catalano, C. Daraio, M. Gregori, H. F. Moed & G. Ruocco (Eds), *Proceedings of the 17th Conference of the International Society for Scientometrics and Informetrics* (Vol. II, pp. 1888–1893). Rome: Edizioni Efesto.

Zuccala, A., Zhang, H. H. & Ye, F. Y. (2019). Mapping disciplinary knowledge flows using book reviews. In G. Catalano, C. Daraio, M. Gregori, H. F. Moed & G. Ruocco (Eds), *Proceedings of the 17th Conference of the International Society for Scientometrics and Informetrics* (Vol. I, pp. 643–654). Rome: Edizioni Efesto.

Zyoud, S. e. H., Sweileh, W. M., Awang, R. & Al-Jabi, S. W. (2018). Global trends in research related to social media in psychology: Mapping and bibliometric analysis. *International Journal of Mental Health System*, *12*. https://doi.org/10.1186/s13033-13018-10182-13036

5

DATA

Sources, processing and analysis

Introduction

In order to accomplish a successful scientometric research, reliable data is a pre-requisite. The data should be available not in a small measure, but in a size that makes a detailed analysis feasible with the applications of statistical procedures. The availability of the required data that meets the intended objectives of a study determines its successful completion. This is why the source of data is so important in scientometrics.

The creation of large electronic databases of scientometric information has been a very significant step in the development of scientometrics. Databases, according to Hood and Wilson (2003), contribute to the provision of data sources and a delivery mechanism or platform through a set of analytical tools. While scientometric data can provide valuable insights into the scientific system, as Sugimoto and Larivière (2018) note, it must be understood within its unique context. This means that an understanding of both the properties of the data and the larger sociological and economic process that govern the scientific system is necessary (Sugimoto and Larivière, 2018) in the selection of appropriate datasets.

Data for quantitative and qualitative scientometrics

Glänzel and Schoepflin (1994) argue for scientometricians to recognise several methods and models that can be applied in different contexts. Valuable concepts that have been developed in neighbouring fields should also be adapted (Glänzel and Schoepflin, 1994). Mainly viewed and used as a quantitative method, scientometrics is statistics-compliant. As with several other quantitative methods, scientometrics makes use of the advantages of statistical tests and procedures. Data can be examined by applying both descriptive and advanced inferential statistical procedures. With scientometric data, researchers are able to formulate hypotheses and statistically test their validity. Correlation tests are easy to run with scientometric data, provided the data is captured and transformed correctly into an appropriate software program. The approach should be creative and innovative not only in the application of methods, but also in the use of scientometric data. The scope of data analysis is, however, not restricted to quantitative measures and procedures. It is just one aspect of scientometrics. De Bellis (2009) realises that scientometrics is not completely citation-addicted nor is it a by-product of a normative theory of science. She is convinced that scientists produce textual material and its surrogates, such as titles, abstracts and keywords, which can be analysed and measured. Exploration of this nature into the textual material, she believes, can detect the associations of scientific concepts for delineating subject areas, subject fields and disciplinary patterns. Co-word analysis is one such technique for mining scientometric text.

Scientometrics can successfully employ qualitative data analysis, but this has not been sufficiently exploited in scientometric studies. Hardly any serious attempts have been undertaken to generate qualitative data. Full-text publications, for instance, are a potent source of data for scientometric research (Lamers et al., 2019). Sourcing of qualitative data from publications, databases and repositories is not difficult to accomplish. This potential of scientometrics is yet to be tapped. Although the literature mentions the scope for qualitative data analysis along with quantitative data analysis, not much attention has been paid to this end.

The most commonly used data sources for qualitative data are people, organisations, institutions, entities, texts (published or unpublished), settings and environments, objects, artefacts, media products, events and happenings (Mason, 2002). For scientometric studies, textual documents are the source of qualitative data. While textual analyses of publications have been undertaken in scientometric studies, they have not reached the level of real qualitative data analysis. Citation indexes, journal publications and books hold a large amount of qualitative textual data. The major parts of qualitative information are in the form of titles of publications, keywords, abstracts and cited references. Metadata of publication records contains all this information. Full-text publications from databases or from journal websites are also a rich source of qualitative data for scientometric studies. Where these are not available, cited references are a source.

Publication records, the units of scientometric studies, contain detailed and descriptive abstracts and keywords. Such rich descriptive information sources are appropriate for qualitative data analysis. Abstracts of publication records, titles of publications or keywords can be processed using qualitative data management software programs such as NVivo, ATLAS.ti and others. Textual analysis of keywords, titles and abstracts unravels the dynamics of knowledge production and communication. This could pave the way for isolating new emerging areas of research, specialities, concepts, terminologies and methodological improvements. Themes and codes can be inferred from this descriptive data which can supplement the quantitative information that has already been captured and processed. These are critical for the development of scientometrics in both science and the HSS.

Lamers et al. (2019) tried full-text publications for the analysis of citation label styles and citing sentences. Full texts are even instrumental in developing metrics for research evaluation (Herrmannova and Knoth, 2014, 2015). Zavaraqi and Fadaie (2012), in one of the rare works, argue for the need to apply mixed methods in scientometrics. They believe that the co-citation analysis used in scientometrics possesses the features of a qualitative method. Actor-network theorists have employed qualitative data analysis in scientometric studies. This is done by focusing on the way in which an author of a publication indexes a bibliographic reference within a line of argument (Gómez-Morales, 2015). A publication may be thus reduced to a network of powerful words and represented diagrammatically.

Imagine that a researcher is interested in studying scientific research conducted in the broad subject area of medicine. Once the data from the database is collected, processed and captured in a suitable software program for analysis, findings of the research in the said area are arrived at. These may refer to the trends of knowledge produced in that subject area, the background of authors (in terms of their affiliation and institution), country or citations. If this analysis is extended to cover the titles, keywords and abstracts or full text of those documents, additional information about subject areas may be revealed. This additional information is actually qualitative in character.

Ordinarily, an abstract contains 150–250 words, depending on the formats of the publication outlets. If hundreds of records are to be examined, which is normal in scientometric research, the wealth of the qualitative data that is hidden in these abstracts is enormous. From just 100 abstracts, 15,000–25,000 words, or roughly 60–100 pages of descriptive data, can be obtained. It is more or less equivalent to the data that can be generated through six to ten face-to-face interviews, each of one hour duration. The titles of publications can also be analysed in the same way, and can provide further descriptive information. Social scientists will find this a treasure trove of information. When researchers have the facility to combine both quantitative and qualitative data, it is easy for them to go deeper into the data. This is actually a mixed methods approach, which gives more credibility to the findings derived from scientometric data. The presence of both quantitative and qualitative information in the publication records makes scientometrics a more acceptable method for researchers who are skilled in this type of research.

The law of Zipf, as discussed in Chapter 1, is applied more fruitfully to the qualitative data that is generated from publication records. The opportunities for qualitative data analysis are therefore not limited. In this chapter, the discussion centres on the major sources of scientometric data, their reliability, challenges and limitations.

Citation indexes

The basic assumption of scientometrics is that the importance of a publication rests with the importance it receives in the form of citations. Normally, a citation index contains a list of references that are ordered in which each reference is followed by a list of sources or citing works which cite it (Garfield, 1964). Citation indexes capture these citations which make scientometric analyses possible. Indexes also provide the data for scholars to test the validity of many scientometric theories (Schnell, 2018). Citations point to the connections between authors, groups of researchers, topics of study, subject areas and impact (Andrés, 2009). The relationships between these can communicate many more complex and unknown dimensions of publications.

Academic communities find citation indexes an indispensable and valuable reservoir of information. The information stored in indexes is crucial to the knowledge of scientific research and its impact on individuals, institutions and countries or society at large. Policy makers and academic and research administrators need this data to formulate their scientific policies. Scientists look to these sources for mapping subjects and disciplines, and for tracing the trajectories of decline, stagnation or growth in specific areas. New subject areas are emerging from these citation sources. Scholars get to know about the usefulness of the references cited in scientific publications that shed light on the research practices, and the social dynamics of scientific knowledge (Gingras, 2016).

Origin, Development and Uses

The entry of citation indexes accelerated the growth and development of scientometrics as an empirical method of study. Indexes are instrumental in the emergence of scientometrics as a specific scientific field, suitable for scientific analyses of scientific literature. Immense opportunities are hidden in citation indexes. In 1955, Eugene Garfield established the Institute for Scientific Information (ISI), which produced the *Science Citation Index* (*SCI*) in 1964. Within a few years, the *SCI* turned out to be one of the few citation indexes that scholars found useful, reliable and beneficial to serious scientometric undertakings.

The *SCI* was not the first citation index to be released. Indexes including *Quarterly Cumulative Index to Current Medical Literature*, *Index Medicus*, *Chemical Abstracts* and *Current Contents* were available then. Citation indexing flourished in the legal world before they were applied in science (Shapiro, 1992). Even Garfield's *SCI* was inspired by the legal indexes, namely *Shepard's Citations*. Frank Shepard began to print citations to Illinois Supreme Court cases in 1873. Alan Pritchard, who first defined bibliometrics, was counting the chemistry publications of 1874 (Shapiro, 1992).

A citation index, as Garfield (1955) envisaged, should have a complete listing of all periodicals covered, and all articles that were referred to in the article. By doing so, the index will provide a complete listing of all the original articles that had been referred to in the article. Garfield believed that such a system of indexing would be useful for historical research. Also, by referring to the listings of one's article, an author could determine which other scientists are referring to their own work, and thereby increase communication possibilities between scientists (Garfield, 1955). Often, this led to future collaborative research.

The *SCI* was conceived to be a tool to facilitate the dissemination and retrieval of scientific literature for the measurement of scientific productivity. The databases of the *SCI* basically serve the purposes of gathering what scientists published, and where and how often these papers are cited (Garfield, 2007). The online platform of *SCI* databases permits users to search and work around the data on publications and their citations. Since its launch, the *SCI* has produced a series of by-products such as the *Journal Citation Reports* and journal impact factor (JIF) rankings that complement indexes. The value of these products is indispensable for the study of scientific aspects of the literature, varying from citations of publications to JIFs.

Historically, citation indexes have been adapted chiefly for the evaluation and impact of scholarly research. In the evaluation of scientific productivity, indexes have implications not only for authors, but also for institutions and countries (Andrés, 2009). Institutions and research funders consider citation analysis a reliable method when making funding decisions. While this was not often explicit in the policies of funders, the relation between the standing of scientists measured on the basis of citations and successful funding applications has been reported.[1] This was actually the expectation: to use the *SCI* as a tool for research assessment, when it was originally introduced.[2]

Garfield admits that he inadvertently contributed to the notion that research assessment is the main *raison d'être* of citation indexes. In his view, it is the only one of the many sociological applications that some scientists know about, and that it is first and foremost a tool for information retrieval and literature search (Garfield, 1983). It means scientometrics has yet to explore the full potential of citation indexes. Garfield has hinted at the underuse of the resources for scientometrics purposes. In his essay, "How to use *Science Citation Index (SCI)*" (1983), Garfield elaborated on the diverse uses of the index that have not been attempted completely. The primary purpose of the *SCI* is to help in finding the published literature on any subject, which is not restricted to the natural or social sciences. Garfield (1984) emphasises the importance of the social sciences in providing answers to policy issues in the physical sciences. The focus of scientometrics also turning towards social sciences and humanities was soon to be witnessed.

Within nine years of publishing the *SCI*, another citation index was to be released by the ISI. In 1973, the *Social Sciences Citation Index (SSCI)* was launched. Until then, the studies that relied on citation indexes were limited only to science and scientific disciplines. It was not only the study of science using the *SCI* but also the study of the social sciences that thus became feasible. The *SSCI* gave a major boost to scientometrics, particularly to those who also wanted to study social science disciplines. The *SSCI* was well received with heightened enthusiasm and interest for its unique features and search functions. Garfield (1975) announced that it is the first and only multidisciplinary citation index for the social sciences.

The *SSCI* embraced all areas of the social sciences, ranging from anthropology to law and to history.[3] The launch of the *SSCI* was a great step forward for the social sciences and for social scientists to know and study their own disciplines and their current trends and development. Its presence caused a change in the approach of scholars, as the study of social sciences was now made possible, like the study of science. True to the definition of scientometrics, as discussed in Chapter 1, the study of both science and the HSS has now been made feasible with a citation index like the *SSCI*. It was an added advantage to use the *SSCI*, as the subjects are linked to each other and to the natural sciences that are indexed in *SSCI*; and this had great importance for research in the social sciences (Garfield, 1975).

Further developments in the provision of data on more disciplines and subjects were to occur in the next few years. In 1978, the ISI published the *Arts & Humanities Citation Index (A&HCI)*. It was yet another additional resource that extended the scope of scientometrics to new domains with additional disciplines and subjects. The *A&HCI* filled a long-felt gap for the humanities disciplines, and includes areas ranging from archaeology to theology.[4] The *A&HCI* accommodated the differences in the format and content of the journals in the humanities, as they differ from science journals, and modified the way the index was prepared for its end users (Garfield, 1985).

In 1988, *SCI* data was released in a CD-ROM format, making it more accessible and usable for scholars. Developments in the creation of new indexes and their content were to continue. The ISI sustained its production of indexes to serve various purposes of the scientific community. It released its Web of Science Core Collection (WoSCC) in 2002, which became a more integrated research platform. In 2015, the ISI presented its *Emerging Sources Citation Index (ESCI)*, which carried journal titles in emerging disciplines and from the emerging world regions.

The indexes are recognised tools of study for the history, evolution and sociology of science by information scientists and sociologists (Garfield, 1984), but much is to be desired in their use by the scientific community. There is a sustained need to educate scholars in the use of indexes (Garfield, 1984). This, as reiterated in this book, is actually a need to attract more people to scientometrics.

Contents

Citation indexes carry a variety of documents. Among them are journal articles, books, book chapters, book reviews, corrections, dance performance reviews, data papers, database reviews, discussions, editorial materials, excerpts, fiction, creative prose, film reviews, hardware reviews, letters, meeting abstracts and summaries, music performance reviews, music scores and reviews, news items, poetry, proceedings papers, record reviews, retracted publications, scripts, software reviews, TV and radio reviews and theatre reviews. The metadata of these documents in indexes contains the topic, title of the paper and publication, author/editor, year of publication, affiliation address of authors (organisation, city, province/state, country/region), funding, times cited, number of references and research areas.

Data from indexes

Scientometric methods are primarily used in citation analysis to find the relationships between journal articles. This is done by examining the references cited in the publications. Scientometrics is also employed to assess the quality of publications. The data obtained from citation indexes is helpful in this regard, as it can be used to assess the impact of research at the individual, institutional and national levels. In recognising the emerging (or disappearing) research areas, specialities, technologies, structure of disciplines and fields, data from citation indexes is useful (Garfield, 1992). As Ball (2018) summarises, the data can yield insights into the key components for a macro research. Among others things, the data is useful for the study of the structure of academic activities in individual disciplines; academic productivity; influence of countries or regions on knowledge domains; collaboration; developments of individual branches of academia; and scientific products and their influence (Ball, 2018).

Scientometric data from citation indexes is of great consequence to the study of scientific production over a period of time across countries, regions or institutions. Without the support of this scientometric data, researchers will fail to track the changing research trends or the emerging fields of research in their area of specialisation (Gingras, 2016). Like many others, Small (2006) used the indexes to track and predict growth areas.[5]

Knowledge about disciplines is nevertheless unavoidable for their own paths towards growth and development. Information on making informed decisions regarding which research is relevant or irrelevant, and what directions a discipline should take, can be deduced from citation indexes. If scholars lack such information, it would be at the peril of a discipline. They need to know about their own disciplines and subject areas, which will also support their own academic careers. For a creative and innovative scholar who wishes to seek and explore new areas of research, scientometric information from indexes is of immense value. This often helps scientists with their ongoing research and to make breakthroughs in their disciplines. In order to develop a new generation of scientists and scholars at both graduate and post-doctoral levels, and to contribute to new knowledge that leads to ground-breaking innovations, scholars seek scientometric data stored in citation indexes.

Scientometric data, stored in databases of WoS and Scopus, is a resource for the study of the social and conceptual changes of science in several disciplines (Gingras, 2016). Undoubtedly, the development of science can also be studied from a historical, sociological, economic or political angle, and on a different scale from micro to macro (Gingras, 2016). With the help of this data, scholars are able to map scientific domains, encompassing a range of aspects including growth, diffusion, specialisation, collaboration, impact, productivity of individuals, groups and institutions, and even obsolescence of scientific concepts and changes in the mode of knowledge production (Björneborn and Ingwersen, 2004; Blakeman, 2018; Martin, 2011). Information available from indexes is applied to locate new areas of knowledge as well. Through scientometric analyses it is easy to identify the emerging areas of research (for instance, Glänzel, 2012; Wang and Chai, 2018), as detailed earlier. Researchers have also successfully utilised indexes to document the works of scholars themselves (Sangam and Savanur, 2010).

More new uses

Many more uses and applications are hidden in citation indexes and are therefore applicable to scientometrics. Owing to the nature of the data, citation indexes are effective for the evaluation of science and technology. Indexes serve not only the purpose of citation analysis, but also assist in developing a set of useful scientific indicators. They have the potential to develop new quantitative indicators as the indexes are multidisciplinary, representing all fields in both science and the social sciences (Garfield and Welljarns-Dorof, 1992).

Institutions rely more on the metrics of impact for hiring able and competent staff for their faculties and departments. The data on authors makes possible an objective and measurable analysis of the visibility and impact of the works of candidates who are looking for positions in universities and research institutions. Career advancement is often determined by the impact of one's research and publications.

Network analyses bear extensive possibilities with scientometric data sourced from citation indexes. It is a fruitful analytical method to inquire into the complex collaborative patterns of authors (De Bellis, 2009). Information on the number of authors, institutions, departments, subject areas, countries and regions indicate the inherent research networks of authors. The networks of scientific collaboration can be reliably marked by the coauthorship networks of authors using scientometric data (Glänzel and Schubert, 2004).[6] Indispensable to the study of authors and their collaboration, network analyses untangle the relations between authors, disciplines, institutions, countries and regions. The networks of scholars, evident from their collaborative preferences, throw light on scientific relations between countries and institutions that have strengths and weaknesses in certain disciplines and subject areas.

Spatial scientometrics is functional to the study of the spatial distribution of research (subjects, disciplines or fields), researchers, collaboration and citations that can be done with information from citation indexes. This kind of analysis can explain the places or locations (regions, institutions or countries) of publications and their distribution across selected parameters. Spatial analyses may be further extended to the study of citations and their location (institutions, countries or regions), researchers (highly cited, lowly cited or not cited) and to institutions, countries or regions. They also lead to an understanding of the internationalisation of publication outputs, focusing on subjects, disciplines, publication outlets and citations.

While performing spatial analysis of publications, researchers and citations, one should be aware of the reliability of information that is used. Like several others, Frenken et al. (2009) are cautious about the quality of information provided in citation databases. The information provided in the affiliation addresses of authors may not always be accurate. If the affiliation addresses are not linked correctly to authors, this will give rise to erroneous results. Multiple affiliations are common in research, which should be taken into account. Adopting Boschma's (2005) proximity on collaboration, Frenken et al. (2009) recommend a proximity approach to spatial scientometrics. The proximity approach should consider the physical, cognitive, social, organisational and institutional dimensions of proximity.

Even in historical research, citation indexes are handy tools. Garfield developed this method with the support of citation indexes. Historical research into citations can be accomplished through a series of steps. De Bellis (2009) explains this in the following stages.

Firstly, choose an essay supplying authoritative and historiographic reconstruction of a significant episode in the history of science. This episode may be a discovery resulting from the accumulation of several previous but small achievements. Draw a map with nodes and lines for the key events and relationships spotted by historians. A literature search is then performed for the nodal papers and investigators who first reported the key events and relationships. In the following stage, the bibliography of each nodal paper is examined to find out the connections with other nodal papers that will lead to a second map. This is what Garfield called the historiography, which shows the bibliographic networks of nodal papers. The two maps thus drawn are then superimposed to analyse overlapping. Lastly, a thorough citation analysis of each item is performed to evaluate the citation impact and to check whether any new connections have been left out by historians.

Joining with Alexander Pudovkin and Vladimir Istomin, Garfield developed a software, HistCite, for historiography to evaluate the output of topical and citation-based searches (Garfield et al., 2002; Garfield and Pudovkin, 2004). HistCite is meant to generate chronological maps of collections from citation indexes. Files with all cited references from source can be created for processing to generate tables of the most cited works (Garfield et al., 2002). The purpose of this is the automatic and interactive construction of historiographs, showing the historical development of a topic (De Bellis, 2009). The software can run tables and graphs that contain chronological sorts, and the frequency rankings by author and journal. The program also holds features meant for vocabulary analyses.

Quality of data from indexes

Although citation data from citation indexes is a unique tool for scientometric studies, its responsible application requires careful and informed interpretation by experts (Garfield, 1992). The quality of data, as for any research, should be a primary concern for scholars who are conducting scientometric studies. Like in any other research in the HSS, the integrity of the data is crucial. The reliability, consistency and accuracy of data (Andrés, 2009) obtained from the indexes need to be checked before it is adapted for scientometric analyses. The quality of the information retrieved from any data source is to be maintained and proper data cleansing routines are to be followed (De Bellis, 2009). An erroneous dataset created from inaccurate records does not hold good for valid and credible scientometric research. Most often, the source of data comes from indexes. Before an index is selected for scientometric studies, it is necessary to assess whether the coverage of the index is suitable to meet the objectives of the research. The coverage, as opposed to the research objectives, should not be biased towards certain countries, languages, publishers or types of document (Andrés, 2009). Multidisciplinary indexes are useful when there are challenges with regard to the coverage of publications. As Larivière et al. (2006) recommend, a number of indexes of WoS such as the *SSI* and *AHSI* should be combined for performing analyses on citation data. If the index is limited in its coverage of journals in regard to the subject, institution and countries, the conclusions derived from the study will have only a limited value.

Combining index data

By now it is clear that with the introduction of citation indexes, scientometrics was in a revolutionary phase of its development (Bornmann, 2013). While indexes are rich sources of information stored from scientific literature produced by publishing houses, they are no longer the only source. Academic material can be drawn from the internet (such as from open access publications, contents of institutional repositories and web pages) as well (Ball, 2018). Scientometric data can also be obtained from sources such as journals and books. Merigó et al. (2018) made a general overview of the trends in the journal *Information Sciences*, using a variety of indicators such as bibliographic coupling, citation and co-citation analysis, coauthorship and co-occurrence of keywords.

Additional and supplementary data combined with citation data from indexes is often rewarding. A combination of data adds more value to the findings as they complement each other, filling the gaps in each set of data. It is common for scholars to be curious to study other pertinent factors and their relationships between key variables. For instance, race and gender of authors are important variables in the study of the productivity and publication patterns. Unfortunately, this information is not saved in indexes such as WoS and Scopus. Such information is not there in publications either. However, it is not difficult to collect this information from other sources such as the personal or institutional websites of authors, or even directly from authors. Nevertheless, it may not be realistic for a macro-level study consisting of thousands of publications. Studies conducted by Aaltojärvi et al. (2008), Aksnes et al. (2019), Leahey (2006), Larivière et al. (2013), Ruggunan and Sooryamoorthy (2016, 2019), Science-Metrix (2018), Sooryamoorthy (2015, 2016) and several others have incorporated additional information on race and gender into their analyses.

Major databases for scientometric studies

A survey of published scientometric studies suggests that scholars, to meet their data requirements for scientometric analyses, often turn to two major sources, namely WoSCC owned by Clarivate Analytics and Scopus of Elsevier.[7] These are the two widely used data sources. The majority of the publications that are carried in prominent journals and other forms of communication are based on the datasets sourced from these two. The content of these data resources is controlled and well defined (Gingras, 2016), which makes the analysis robust to produce valid findings. On their respective online platforms, effective tools are embedded for preliminary analysis. While these tools are not adequate for deeper levels of analysis, they serve the purpose of preliminary descriptive analyses. Google Scholar (GS) is another source of scientometric data, but it is not as prominent as the major ones owing to several limitations including the questionable quality of data. The content of GS is not well defined and varies significantly (Gingras, 2016). Particularly for large-scale analyses, GS poses challenges (Visser et al., 2019). GS has, however, been used as an experimental start in comparison to and in conjunction with other major databases. A detailed description of these data sources follows.

Web of science

The WoSCC of Clarivate Analytics (Figure 5.1) has six databases: the *SCI*, the *SSCI*, the *A&HCI*, the *Conference Proceedings Citation Index*, the *Book Citation Index (BkCI)* and the *ESCI*. This core collection covers the publication records of over 252 disciplines from as far back as 1900. As of October 2018, WoSCC covered 20,936 journals, 94,066 books and 197,792 conference titles that included 10,443,486 records.[8] Access to these is controlled by subscription. Depending on the institutional or individual subscription, access to its different databases varies.

The *SCI Expanded* now covers over 9,046 major journals in 150 disciplines starting from 1900. Currently, the *SSCI* has over 3,330 journals across 55 social science disciplines and selected items from 3,500 scientific and technical journals from 1900 onwards.[9] The *A&HCI* initially indexed 1,815 humanities journals and

Web of
Science
Group   Who we are for   Products   Resources   Services   ISI $\bigcirc$ | A **Clarivate Analytics** company

Find out more about our editorial process

Web of Science Core Collection

A trusted, high-quality collection of journals, books and conference proceedings.

Learn more

Science Citation Index Expanded (SCIE)	Social Sciences Citation Index (SSCI)	Arts & Humanities Citation Index (AHCI)	Emerging Sources Citation Index (ESCI)	Book Citation Index (BKCI)	Conference Proceedings Citation Index (CPCI)

FIGURE 5.1 Web of Science Core Collection

included contents from 250 scientific and social science journals from 1975 to the present. The *ESCI* under its coverage has 7,280 journals published in over 200 countries. Its *BkCI*, which was launched in 2011, includes both series and non-series types, scholarly books that have fully referenced articles of original research, and reviews of the literature from 2005 onwards. It currently covers over 80,000 books and 150 highly cited book series in science, social science and humanities. The *BkCI* has two subindexes, *Book Citation Index – Science* (*BkCI-S*) and *Book Citation Index – Social Sciences and Humanities* (*BkCI-SSH*). Approximately, 10,000 books are added to this collection every year. Similar inclusion criteria as in other WoS databases are used (Adams and Testa, 2011). More and more applications of the index are coming through (Glänzel et al., 2016), but it is not free from coverage issues. Its biased coverage and incomplete records have been reported. Also, it is biased towards books published in the USA, major publishers in STEM and in English (Sugimoto and Larivière, 2018). The *Conference Proceedings Citation Index* covers over 180,000 conference proceedings from 1990 to the present.

WoS has a consistent editorial process for the inclusion and exclusion of titles in its databases. It chooses the world's best journal titles and the criteria of inclusion and exclusion are influenced by the Bradford and Garfield laws of concentration, which state that the core literature for all scientific disciplines is concentrated within a small set of core journals (Schnell, 2018). The information saved in the citation indexes is created through a careful and sophisticated system of entering, checking, editing and verifying. Every entry is scrutinised by a combination of both human and computer editing, and is then machine verified. WoS adopts both computerised and human indexing tools to capture journal issues from cover to cover (Schnell, 2018). Titles in other languages are translated into English (Garfield, 1975). Often, only the journals that receive a particular impact factor over a longer duration of time are evaluated in WoS (Ball, 2018). The approach to selection therefore depends on the quality and impact of the journals. WoS covers the best literature, which is about 10 to 12 per cent of all the journals published worldwide. For journal publications, information on all authors, their affiliations, abstracts, keywords, funding details and all cited references are captured.

The *SSCI* serves many purposes for its users. Researchers can trace the development of thought in the social sciences by examining the traditional disciplinary histories, and tracing the impact of one's work (Garfield, 1975). It also allows users to find the kind of research that has systematic repercussions with the production of publications in the field, as well as the dead-end research in which publications have failed to influence further investigation (Garfield, 1975). Some of these uses have been described in Chapter 4.

Subject categories in WoS for both the *SSCI* and *A&HSS*, with a description of their respective coverages, are given on their websites. While choosing a particular subject category for scientometric analysis, the coverage of the categories has to be taken into account.

Despite this, WoS is criticised for the extent of coverage of journals. Only a selected set of journals is part of the WoS collection. The journals in the collection also tend to be the highest impact peer-reviewed ones and therefore represent only a fraction of research done in any given field (Neuhaus and Daniel, 2008). The bias of WoS towards English-language journals from English-speaking countries is its drawback. Studies (Archambault et al., 2006, for instance) report that journals with editors in countries where the language spoken is English or Russian are overrepresented in WoS databases. Meanwhile, languages such as French, German and Spanish are underrepresented in the database (Archambault et al., 2006).

The new *BKCI* of WoS (with its two editions, *BKCI-S* and *BKCI-SSH*) is deemed to be a strong step in the direction of creating a database for monographs, which is more beneficial to the HSS. The content of a book is examined by applying certain standards for inclusion in the database. In order to qualify for inclusion in the databases, a book should be a full-text original research (print or e-book), have a bibliography, references or footnotes, and a copyright year dated 2005 or later (as the *BKCI* indexes books published in 2005 and later). Books based on maps, atlases, textbooks or abstracts are omitted (Gorraiz et al., 2013). It is, however, fraught with limitations. The *BKCI* has a high share of publications without the address information of authors and inconsistencies in citation counts (Gorraiz et al., 2013). These issues might be resolved in the near future so that an additional source of data for the HSS is possible. For the HSS, monographs and edited books are indispensable sources of information.

Scopus

A recent entrant to the field of citation databases, Scopus was launched in 2004 (Figure 5.2). Since its inception, the database, as noted in Chapter 1, has grown in its coverage and storage of citation data. Scopus is named after the African hamerkop (*Scopus umbrette*), a medium-sized wading bird, which is famous for making huge nests as wide as 1.5 metres (Schotten et al., 2018). The metadata of journals,

FIGURE 5.2 The opening search page of Scopus

conference proceedings and scholarly books and their citation content is preserved in Scopus.

Scopus has functionalities of author profiles and author affiliation profiles. Its Scopus Citation Tracker enhances citation analysis by viewing citations year on year (Neuhaus and Daniel, 2008). The built-in feature of author disambiguation automatically assigns publications to authors with a high level of precision (Sugimoto and Larivière, 2018). A committee is responsible for the acceptance, selection and inclusion of journals in Scopus. Journals indexed in the database are also subjected to re-evaluation. As scientists and editors can recommend journals to Scopus, this process helps journals with a lower JIF to get accepted as part of the database (Ball, 2018). The committee is open to publications in languages other than English, which adds to its broader coverage. The database also has a strong European and Asian focus (Ball, 2018). Scopus has an advantage over its competitor WoS as it owns the material it indexes, whereas WoS receives information from publishers about publications from which metadata is extracted.

Google Scholar

Launched in 2004, GS is a freely accessible search engine for scientific literature, covering a list of disciplines, document types and languages (López-Cózar et al., 2018). GS covers journals, books, conference proceedings, dissertations, reports, pre-prints, post-prints and other scholarly documents. These are sourced from academic publishers, pre-print and post-print servers, bibliographic databases, repositories from universities, research organisations and government agencies (Neuhaus and Daniel, 2008). Two of its products are relevant for scientometric studies: Google Scholar Citations (GSC) and Google Scholar Metrics (GSM).

GSC is a tool introduced in 2011 to create an academic profile of a scholar, taking into account all contributions of the scholar that are indexed in GS. In the academic profile, the total number of citations of the chosen author, the h-index and the $i10$-index, referring to the number of articles which have at least ten citations, can be found. These figures are shown for all years including the last five years. As seen in the snapshot of Robert Merton in Chapter 2 (Figure 2.3), all the citations earned by each of these contributions are displayed on this page. The listed contributions can be merged if there are duplicates, deleted if they are about a different author, or exported to reference management software programs like EndNote. This allows the author to see their contributions and their specific and individual impact on the scientific community. They can also be arranged according to year (new or old). Suggestions by the coauthors of the searched author are compiled on this platform. GSC also offers personalised alerts when new articles, new citations and recommended articles are added to the profile, areas of interests and lists of authors by institution. This makes the impact of other scholars in research areas comparable to that of the author.

GSM is Google's journal-ranking platform introduced in 2012. A snapshot of this page is produced in Figure 5.3. GSM can be accessed from the dropdown menu

≡ Google Scholar

◆ Top publications

Categories ▾   English ▾

	Publication	h5-index	h5-median
1.	Nature	362	542
2.	The New England Journal of Medicine	358	602
3.	Science	345	497
4.	The Lancet	278	417
5.	Chemical Society reviews	256	366
6.	Cell	244	366
7.	Nature Communications	240	318
8.	Chemical Reviews	239	373
9.	Journal of the American Chemical Society	236	309
10.	Advanced Materials	235	336

FIGURE 5.3 Journal ranking in Google Scholar

on the top of the screen under the name Google Scholar. A click on the three lines brings up four options, namely My profile, My library, Metrics and Alerts. The Metrics option goes to the "Top publications" pages as shown in Figure 5.3. Top-ranking journals are listed according to categories of subject area, the $h5$-index (h-index for articles published in the last five complete years) and $h5$-median (the median number of citations for the articles that make up the $h5$-index). Under the Categories menu, there are business, economics and management; chemical and material science; engineering and computer science; health and medical sciences; humanities, literature and arts; life sciences and earth sciences; physics and mathematics; and social sciences. The journals can also be chosen according to 12 languages (English, Chinese, Portuguese, Spanish, German, Russian, French, Japanese, Korean, Polish, Ukrainian and Indonesian).

GSM can also show the most cited publication in each of the journals ranked on its platform. In Figure 5.3, the top ten journals are shown. When the $h5$-index (362) of the journal *Nature* is clicked, it will open the page where articles are arranged in descending order of the number of citations (Figure 5.4). The article "Deep learning" in the journal *Nature* has the highest number of citations (8,519). When this number is clicked the link opens to all the references that cited this paper. The results are released on a separate page.

GS has its own share of challenges in regard to the standardisation and reproducibility of the data, but the data it provides has been improved recently. Winter et al. (2014) analysed the development of citation counts in both WoS and GS, and found that GS has expanded substantially and the majority of the current works indexed in WoS are also found in GS. Despite its advantages, scholars have noted its inability in citation studies. Aguillo (2012) held that GS lacks the quality control required for its use as a scientometric tool and its coverage is not comparable with

← Nature

h5-index:362   h5-median:542

#1 Life Sciences & Earth Sciences
#1 Life Sciences & Earth Sciences (general)

Title / Author	Cited by	Year
Deep learning. Y LeCun, Y Bengio, G Hinton Nature 521 (7553), 436-444	8519	2015
Sequential deposition as a route to high-performance perovskite-sensitized solar cells. J Burschka, N Pellet, SJ Moon, R Humphry-Baker, P Gao, ... Nature 499 (7458), 316-319	5377	2013
Efficient planar heterojunction perovskite solar cells by vapour deposition. M Liu, MB Johnston, HJ Snaith Nature 501 (7467), 395-398	4391	2013
The global distribution and burden of dengue. S Bhatt, PW Gething, OJ Brady, JP Messina, AW Farlow, CL Moyes, ... Nature 496 (7446), 504-507	3702	2013
Van der Waals heterostructures. AK Geim, IV Grigorieva Nature 499 (7459), 419-425	3651	2013

FIGURE 5.4 Articles in a selected journal and citations in Google Scholar

that of other databases. Gingras (2016) reported that the real content is not clear as it includes both the peer-reviewed papers and other documents that anyone can put on a personal website. Items from this source may appear or disappear at any time, affecting the validity of the analysis that uses this data source (Gingras, 2016).

GS, having a broader source than other competitors notwithstanding, is the most enigmatic database. Jacsó (2011) calls it enigmatic as it does not provide the details of its sources, the number of records collected, the number of journals, conference proceedings and papers, books, book series, theses, patents and presentations. Neither the algorithm behind its content nor its search algorithm is publicised (Sugimoto and Larivière, 2018). GS does not supply clear information about what is and what is not in its database. The records are not usable for systematic and rigorous analysis (Martin et al., 2010). Transparency on the coverage of the information remains an issue for GS as well (Harzing, 2014; Sugimoto and Larivière, 2018).

The products of GS can be easily manipulated. To prove the vulnerability of the data on GS, López–Cózar et al. (2012) conducted an experiment by faking an author. The research profile of the faked author showed how false documents can modify the h-index. Reportedly, GS comes up with a higher number of publications for many authors who may not even be aware of having written them (Jacsó, 2011).

In a bid to study the ability of GS to replace other databases such as WoS and Scopus, Halevi et al. (2017) reviewed some comparative articles. They compared the articles with those in WoS and Scopus to see whether GS could be used as a reliable source of data for scientific information. Comparison was done on the three source levels of coverage, citation tracking and authors. The results indicate that GS has significantly expanded its coverage, but caution should be exercised when using GS for citations and metrics as it can easily be manipulated.

Adriaanse and Rensleigh (2013) compared WoS, Scopus and GS to find the database which is more representative in the coverage of South African environmental sciences citation coverage. WoS retrieved most citation results, then GS and lastly Scopus. Both WoS and Scopus did not retrieve duplicates, but GS did. GS also retrieved the most inconsistencies in content verification and content quality, including author spelling and sequence, volume and issue numbers (Adriaanse and Rensleigh, 2013). Harzing and Wal (2008) argue that GS has additional advantages over other products, such as JIF, and its free availability allows for democratisation of citation analyses.

The limited search capabilities of GS continue to be an issue for more complex analysis. The data cannot be easily downloaded or analysed (Halevi et al., 2017). Data is collected via its search engines and not directly, as in WoS and Scopus, which poses its own problems. Bar-Ilan (2001) warns that data collection from the web must be carried out with great care, as the web constantly changes and the quality and reliability of most of the search tools are not up to scratch. The poor quality of data and the lack of critical information about the authors' institutional details do not make GS preferable to WoS or Scopus.

After conducting a content analysis to check the accuracy, relevance and applicability of Google search results, Holmberg and Bowman (2019) reported that while GS contains a vast amount of data, it does not come without problems for altmetric research. This is mainly because GS omits some search results, as it judges them to contain duplicate contents of a page already shown in the search results.

A free software program, Publish or Perish, developed by Anne-Wil K. Harzing in 2007, is capable of retrieving and analysing citation data from GS (https://harzing. com/resources/publish-or-perish). The program can supply data and metrics on the number of papers, citations, the average number of citations per paper, per author and per year, the average number of papers per author, the h-index, the g-index, the e-index and the age-weighted citation rate. But the software is not free from limitations. It cannot process more advanced scientometric analyses including global, institutional or disciplinary analyses (Sugimoto and Larivière, 2018).

Comparing WoS, Scopus and GS

WoS, Scopus and GS are the three major databases used for scientometric studies, but they are not the same. The comparative study of WoS and Scopus by Visser et al. (2019) reveals that Scopus covers a large number of documents that are not covered by WoS; all journal articles covered by WoS are also covered by Scopus; WoS covers meeting abstracts and book reviews which are not in Scopus; and a substantial share of the proceedings papers are in WoS and not in Scopus. The cardinal difference between GS and the other two databases lies in the types of data they provide. GS, as against WoS and Scopus, does not collect data directly, but searches the web (Mingers and Leydesdorff, 2015).

Comparing the three data sources, Jacsó (2005) summarises their features and functionalities. Both WoS and Scopus have powerful features for browsing, searching, sorting and saving. In contrast, this is not the case with GS, which has only limited search and sort functions. Both WoS and Scopus have well-facilitated features for scientometric searches, which are non-existent in GS (Jacsó, 2005). GS does not supply information on the source of data, document types, time ranges or update frequencies (Bornmann et al., 2016). It is not an alternative to WoS or Scopus as it is not clear which publications are accepted or how the citations are gauged (Ball, 2018). The structure of the GS platform does not provide organised information on several crucial indicators (Gingras, 2016). The addresses and countries of all authors, which are vital sources of information for scientometric analyses, cannot be viewed on the GS platform. Similarly, the list of references in the papers and classification into subfields are also lacking in this source (Gingras, 2016). Undoubtedly, these are very limiting for comprehensive scientometric analyses.

Citations in GS are not beyond question. Unlike other databases, GS indexes documents that are not purely scientific. It does not offer adequate metadata of the publications that are given in WoS or in Scopus. In a comparative study of citations in WoS, Scopus and GS, Martín-Martín et al. (2019) revealed that the largest percentage of citations across all areas was found in GS. GS had nearly all of WoS (95%) and Scopus (92%). In GS, most of the citations were from non-journal sources (48–65%) including theses, books, conference papers and unpublished material. Prins et al. (2016), who collated both WoS and GS data, confirmed these differences. Retrieval of data from GS required considerable effort on the part of researchers to ensure the quality of data (Prins et al., 2016). This acts as a deterrent for scientometric research.

Scopus is said to have a broader coverage of the publications in the HSS than WoS, and it has made a commitment to indexing the humanities through the inclusion of national journals (Sugimoto and Larivière, 2018). This is an advantage for Scopus. Given the nature of the data available on the GS platform, complex analysis which is possible with other databases cannot be performed with GS data. It is only suitable for individual-level analysis (Sugimoto and Larivière, 2018).

Comparisons have also been made between WoS, Scopus and other lesser-known citation indexes. Ma and Cleere (2019) refer to the Output-Based Research Support Scheme (OBRSS), which was implemented by the University of Dublin in 2016. In their comparative analysis, OBRSS was found to have a more comprehensive coverage of the HSS disciplines than WoS and Scopus.

Data reliability

The issue of reliable data that can be adopted for scientometric analyses appears quite frequently in the literature. Scientometric studies need data in sufficient quantity and quality. The lack of reliable and adequate data limits the study of any discipline in the HSS. Bornmann and Leydesdorff (2014) caution against some of the limitations on available and accessible data. Primarily, scientometrics is applicable only to disciplines in which literature and citations from appropriate databases are available and accessible. The distribution of numerical data in the database is another issue that Bornmann and Leydesdorff wanted researchers to be aware of. Owing to some having highly cited or no cited publications, citation data may be skewed and may require further processing with statistical tools. While using citation data, a window period should be carefully chosen as it will influence measurement and findings. A window period of at least three years is generally acceptable (Bornmann and Leydesdorff, 2014). For publications in the HSS, the time taken to earn citations is more than in science publications. For the HSS a longer window period is advised, as citation statistics is discipline dependent. Some disciplines are slow in accumulating citations while others are faster. This relates to the discipline-based citation culture as discussed in Chapter 3.

Finding reliable data

Finding reliable and accurate data is the crucial first step in scientometric studies. If a study is about the HSS, then the appropriate channels of publication of the disciplines in the HSS should be sought. Often, most of the publications in the HSS might appear in the same publication channels. Guns et al. (2018) report that a majority of the publications in the HSS are published in HSS channels.[10] The coverage of publications in respect of disciplines and subject fields is a persistent issue in scientometrics. It is best to seek proper databases or specific journals as the main sources of reliable information.

Nevertheless, it should be kept in mind that the information stored in databases of WoS and Scopus are not sample based, but population based (Sugimoto and Larivière, 2018). Publication documents are selected according to their own set of standards, criteria, quality and impact of the journal. Subject areas in many cases constitute the primary criterion for inclusion, and are not a representative sample of publications in any discipline, language or country.

Despite the limitations of the prominent databases in regard to the coverage of journals, they are still valid sources of data for scientometric research. It is not practical or realistic to study publications of all journals in any given subject or field. Sampling therefore becomes an acceptable norm and practice for both the producers of indexes and researchers. Scientometricians adopt a sample of data by resorting to a proper sampling strategy. The journals that are indexed in databases thus present a sizable sample of publications in the subject areas. The proviso that it represents only part of all publications in the subject area should be adequate to caution the readers about the limitations of generalisations. These limitations can be overcome by taking advantage of other region-based databases and individual journals.

Conceptual clarity of records

It must be reiterated that the quality and reliability of data to be used for scientometric analyses is crucial. As much as the quality of data, the conceptual clarity of publications that are to be used in the analysis is important. One should be clear about what the types of publication mean – journal articles, reviews, monographs, edited volumes, book chapters, conference proceedings, doctoral dissertations, reports and other documents. They are different types of document bearing their own characteristics. Similarly, other concepts and indicators (subjects, fields, institution, country, publisher, coauthorship, year of publication, language, for instance) are to be considered and defined clearly and carefully. Conceptual clarity adds precision to the data used. Documents are available to assist in this regard.[11]

Before a database is chosen, a researcher should be aware of a few issues that can influence the outcome of the study. According to Sugimoto and Larivière (2018), time, data quality, normalisation, coverage and alignment are important issues that will have consequences on the results of the analysis. By time, they mean the publication and citation windows that must be appropriate for the analysis. High-quality works take time to receive and accumulate citations, and citations vary according to disciplines and subjects. A representative time window, as noted earlier, should be carefully considered for any scientometric analysis.

Data quality

Data quality can be compromised only at the expense of the quality of the study. Quality must be ensured in every bit of information that is to be used. Sugimoto and Larivière (2018) discuss the issue of data quality relating to author or institution names. These are inevitable variables in coauthorship, collaboration, citation and impact. Spelling errors, duplication and even missing information are to be checked and cleaned in the preparation phase of a dataset. It is to be maintained that the selected database has adequate coverage of the topic under investigation and, in the case of multiple disciplines, normalisation should be done before the analysis is performed. The issue of alignment refers to the alignment of the objective of the study with the selection of the data. This becomes significant in research evaluation wherein indicators are aligned with the objectives of the study.

Considerations in the selection of data

The selection of an appropriate and reliable database must be based on the considerations of coverage; types of publication document required and available language of publications that are represented in the database; availability of historical data stored in the databases; and the reliability, credibility, consistency and accuracy of data.

Apart from the major databases, scientometric scholars rely on other dependable sources of information. Publications or the metadata of publications from the home page of journals constitute another reliable source of scientometric data. Not only the metadata of publications (except citation data that can be collected separately), but also full-text publications, subject to open access or subscription, can be downloaded from journal sites. Such information is advantageous when similar information about a particular journal or journals is not obtainable from major databases. Metadata of book publications including abstracts may be collected from online bookstores (Amazon and Google Books, for example). Both these kinds of data collection from journal and book sites are time consuming as data has to be generated, as opposed to using data that is already stored in indexes, but they are crucial and vital for some scientometric analyses.

Online access

As most of the prominent databases are now online, the required data can be mined from their websites. The online facility, in contrast to the old printed or CD-ROM versions, is beneficial to researchers as the databases are regularly updated with the latest information. The majority of academic journals now have both print and online versions, and records can be accessed from the web, making mining data easier than ever. Leopold et al. (2004) describe three kinds of data mining from the web: web structure mining, web usage mining and web content mining. What is important for scientometric research is web content mining: drawing useful information from relevant web documents provided on appropriate database platforms.

Concerns, precautions and preparations

Concerns, precautions and preparations in the process of sourcing data are to be taken into account. Decisions taken at every stage determine the meaningful realisation of the research objectives, the integrity of the process, the quality of the analysis and the worth of the findings. To begin with, suitable data sources are to be selected which depend largely on the objectives of the study (citation analysis, research evaluation, mapping or collaboration). Databases are the sources of raw data and have some basic tools for sorting and analysis (Hood and Wilson, 2003). The selected database or other data sources must be capable of providing the requirements of a study. This is not to pre-empt that the databases do not have any problems or challenges at all (Hood and Wilson, 2003).

Decisions are to be made before data is sourced from the chosen database. These pertain to the objectives of the study, the types of document, period of data, institution or country, language and subjects. Documents are available in several forms such as articles, monographs, chapters, conference proceedings, reviews, notes and letters. Depending on the purpose, subject areas and coverage, the most appropriate types of document are to be chosen. They can be journal articles, chapters, books or all of them.

Before the actual collection begins, the database or journals need to be consulted for their coverage. If they do not contain data to meet the objectives of the study, alternate and supplementary sources of information are sought to fill the inadequacies. By opting for another suitable database or supplementing it with other data sources, the inadequacies can be addressed.

Basically, the coverage is to be examined from the standpoint of the inclusion or exclusion of sources (journals, monographs, edited books, conference proceedings, reports, dissertations or other documents); types of document (journal articles, chapters, monographs, reviews, notes or commentaries); subjects or disciplines (which should correspond to the focus of study as some databases have a better coverage of some subjects/disciplines than others); countries, regions or institutions (depending on the purpose of the study); citation details (the number of citations and references cited in the document); language; time span (the period of availability determines the success of historical and trend analysis); and publisher details (by whom and where documents were published).

The accuracy of data stored in the databases should be thoroughly and cautiously checked before it is imported. Databases might have erroneous spellings of authors, institutional affiliations and journal names. Change of journal names is not an unusual phenomenon in academic publishing. By carefully searching the accuracies of suspicious information, these errors can be checked and corrected in the cleaning and processing stage.

Incomplete publication records, which is not abnormal in databases, pose a challenge to scientometric analyses. Remarkable improvements have, however, been made in providing complete information of records, particularly since citation indexes have gone online. The missing fields of information from the publication records are sometimes difficult to cover. The main gap in publication records is when the names of all authors in coauthored publications and their institutional affiliations are missing. For a scientometric study of research collaboration, such incomplete information in a significant measure can be problematic. Like the names of all authors, their institutional affiliations are also necessary when the study is intended to do an institutional analysis of research performance or collaboration. In other words, all bits of information are valuable for a proper and complete scientometric research. Finding the missing information is achievable through searching the internet, if the study takes up a manageable sample. Institutions and authors are now more visible in the public domain than before.

The vintage of data chosen and the period during which data is collected are the next concerns. In any scientometric research, information about the vintage of data cannot be omitted from reporting. Historical data for all disciplines, whether journal articles, monographs or chapters, may not be available or accessible for the selected period. Researchers have to be satisfied with what is available and what can be created. Stipulations that a particular period is the only way to do a scientometric research are impractical. Researchers, editors and peer reviewers are aware of this limitation in scientometric studies. While there are no rigid rules for a particular window period, a three-year window is generally acceptable. It is still contingent on the research objectives, disciplines and subject fields of the study.

The window period or time span for the analysis is determined by several influencing factors of the study. It may vary from a single year to many years. If the analysis is of individuals or institutions, a smaller time span is acceptable. The spread should expand to several years or to some sampled years, when the variables (subject(s), country/countries or the total publication count) demand trend analysis. Multiple years become necessary when a study compares differences between institutions, countries or subjects. A broader window is justified for mapping the historical development of a discipline or subject field. Some have used a longer period of 50 years in their research. The choice of a period of analysis is therefore justified by the above considerations.

In the selection of the period of analysis, the possibility of having a large number of documents might pose a challenge. Thousands of publications are produced in an advanced country in any given year. A study of a huge magnitude may not always be feasible when resources to capture the data into a computer program are meagre. Manageable numbers of records make data processing easier. Since most of the databases do not have the menu for advanced statistical analysis beyond descriptive statistics, researchers have to source the data and recapture it into a computer program. This is a labour–intensive and time-consuming exercise. As in any research project, the researcher has the freedom to choose the appropriate time period which can be justified on methodological grounds.

Sometimes a sample of publications would be adequate. However, the size of the sample is always a concern for scholars when publication records are available in huge numbers. Samples that are either too small or unnecessarily large, as Williams and Bornmann (2016) argue, have disadvantages. One needs to take into account what effect the sample will have on the analysis. In view of the intensive labour involved in downloading, capturing and processing data, a sample from a large slice of publication records is often convenient for scholars undertaking studies. A properly selected sample will not affect the application of statistical tests or the capability of generalisation and prediction of data. Sampling is convenient for researchers to undertake studies at the micro or meso levels of analysis.

Categorisation of documents is another concern. A publication document in a database may be categorised under more than one subject area, depending on the content, scope and relation to other subject areas or disciplines. For instance, it is not rare to find a publication grouped simultaneously under the subject areas of information and library science, communication or computer science. This throws light onto the interdisciplinarity of the publication and this aspect, if need be, can be incorporated into the analysis. It is preferable to find all the related subject areas under which a publication document has been populated in a database.

Databases are always constantly updating with the addition of new metadata being produced. From time to time, metadata is frequently reloaded, reindexed or removed from hosts (Hood and Wilson, 2003). Therefore, the date or period of mining data is consequential. Regular updating of records affects the reproducibility of findings, as an analysis done in two different periods of the same dataset may not produce the same results. Information about the time of data retrieval should be saved safely for inclusion in the reports and publications that emanate from the analysis.

The most important thing to keep in mind while collecting data from databases is to gather all the available information for the set research objectives. If a study is about collaboration, the collection of institutional affiliation of all authors and not just the first author is unavoidable. Andrés (2009) notes that it is advisable to list the affiliations of all authors of the documents, which is important for an accurate analysis. The process is seemingly laborious but provides a better description of author participation and collaboration (Andrés, 2009). Such information is indispensable for a precise analysis of different types of collaboration. When it is a citation study, aspects of citations such as total citations and references cited in the documents are necessary.

The discussion thus far has centred on collecting, downloading and saving data from databases from citation indexes. As indicated earlier, it is beneficial to take advantage of specific and core journals to study disciplinary features and trends. Data obtained directly from journals, as against the metadata from citation indexes, is often found to be supplementary or independent in scientometrics. Like citation indexes, data deduced from publication documents of core disciplinary journals is a valuable resource in scientometrics. Note that since data has to be collected manually, the volume of data to be used for a study cannot be too large like the data collected from citation indexes.

Processing data

Processing and analysing data are the major phases in research. Processing involves the crucial steps of checking, editing, coding and transforming (Kent, 2015). Since data is being transferred from one format to another, i.e., the raw data into a data management program, capturing is part of data processing. This is necessary since a detailed analysis of data cannot be performed with the given tools on the online platforms of databases and journals. After data has passed through the stages of checking, editing, coding, transforming and recreating, analysis follows. Admittedly, there are obvious differences between quantitative and qualitative data in the way they are processed and analysed.

The basic unit of analysis is publication records – journal articles, chapters, books, reviews, conference proceedings and several others as described in previous chapters. The metadata of publications allows for extensive possibilities of analysis. In the analysis the unit of publication should be treated cautiously. Some believe that publications are not independent units, but are dependent upon other factors. For instance, publications are dependent on subject, coauthor, country and institution and therefore they naturally vary. A more complex analysis should take this interdependence into account by even stratifying the sample to reflect different strata in the analysis (Williams and Bornmann, 2016).

Citation indexes take care of the metadata of publication records stored in their respective databases, but this needs to be thoroughly checked before it is ready for analysis. Each publication record has to be checked and verified for accuracy and consistency. The volume of the records that are to be used is no excuse for the researcher to use the data in the way it is available. Inaccuracies and inconsistencies are common occurrences in the spelling of the names of authors and institutions, year of publication, volume and issue numbers, page numbers and citations. Most of these elements of metadata can be checked and cleaned while the records are collected and captured into a software program. It is easy to identify the errors while the metadata of the publications is being captured individually into a data management program. In scientometric studies, the size of the sample is invariably big, thus the support of a purposeful data management program is necessary. The online options in place for several citation indexes can only support basic analysis and do not provide appropriate facilities (Moed, 1988). The stage between the sourcing and the capturing of data into a software program is the most labour-intensive part of the process and demands sufficient time and resources.

Analysing data

After all errors have been identified and corrected, the dataset is now clean to run both descriptive and inferential statistical tests. A good understanding of statistics is expected. Quite a few books on applied statistics are available on the market.

Analysis takes place at different levels, from simple descriptive analysis to complex multivariate analysis and to statistical models. As is usual in any analysis, the basic requirement is to know the right variables for the research questions and objectives. In multivariate analysis, the researcher should carefully choose and consider what variables are to be included in the model (Williams and Bornmann, 2016). When regression models are run, the selection of appropriate variables is to be done meticulously without losing the focus of the study. Only the tests that are really required for the specific objectives of the study need to be used.

The data generated from citation indexes and publications has its own weaknesses and limitations. Most of the distributions in the data are skewed (Glänzel, 2008; Schmoch, 2019). Because of this, one needs to be careful about the application of statistical tests. As mean values are regularly used, skewness in the data should be checked. Several methods are available to deal with this. Schmoch (2019) recommends that standard mean values should be replaced by adjusted mean values, where outliers with very high positive citations are cut while low or no citations are cancelled.

The approach in qualitative data analysis is inductive. Patterns, themes and categories of analysis are extracted from qualitative data. As Srivastava and Hopwood (2009) argue, patterns, themes and categories are driven by a set of factors such as the subscribed theoretical frameworks, subjective perspectives, ontological and epistemological positions, and understandings of the researcher. In qualitative data, the researcher looks for concepts, themes and their connections to study something that is related to them. It is a constant hunt for concepts and themes that will provide the best explanation of "what is going on" in an inquiry (Srivastava and Hopwood, 2009).

Several different approaches and analytical practices are used in qualitative data analysis. One practice which is relevant here is to sort and sift through the material to identify similar phrases, relationships between variables, patterns, themes and sequences (Harding, 2013; Miles and Huberman, 1994). This can be done manually or with the help of computer programs. Computer-assisted qualitative data management programs make qualitative data analysis faster and more efficient as they supply a range of categorisation and reporting capabilities (Blank, 2004). Software packages such as Atlas.ti, Ethnograph and NVivo are available, as are books on how to use them (Lyn, 1999 and Bazeley, 2007, for instance).

Graphical presentation of data

Data can be presented in attractive graphical forms too. Charts and diagrams can be produced with the help of programs such as SPSS and Excel. Specific software programs are convenient for presenting data graphically, particularly in mapping studies. Among several, the one commonly used in scientometric studies is the free software tool VOSviewer (www.vosviewer.com), which is the short form of visualisation of similarities.

VOSviewer, developed in 2010 by Nees Jan van Eck and Ludo Waltman at the Centre for Science and Technology Studies, Leiden University, the Netherlands, is intended to create, visualise and explore maps of subjects. It can be used for analysing different kinds of bibliometric data such as network, citation and textual data. As the developers of the program have described, it has different functionalities (van Eck and Waltman, 2010; van Eck et al., 2010). It can display a map in many ways, emphasising different aspects of the map. It can zoom, scroll and search, and can deal with a large number of items of over 100. Two types of map, distance-based and graphic-based, can be created. In distance-based maps, the distance between two items is shown with the strength of relations. In graphic maps, the distance between two items does not necessarily reflect the strength of relations. Detailed step-by-step procedures are provided in van Eck and Waltman (2014).

The need for theoretical developments

When it comes to the application of theories in scientometric studies, particularly in the HSS, there is much to be desired. Theoretical underpinnings and foundations in research have a large role and significance in a majority of the disciplines in the HSS. The application, empirical testing, modification and development of theories constitute an integral part of research in any discipline. Scientometrics is not an exception to this rule. As for the research conducted in the fields of the HSS, the applications, testing and developments of theories have a greater effect on building disciplinary strengths. By and large, scientometric research is centred on a few theories of Lotka, Bradford, Price, Zipf and Merton. They are strong theorists whose contributions have taken scientometrics to where it is today. No one will disagree with the fact that these scholars laid the theoretical foundations for scientometrics. Despite this historical context, it is imperative that scientometrics grows beyond the fundamental foundations of its fathers, seeking new horizons in the theoretical world.

Although scientometrics is about communication, publications, authors and citations, theories should not be restricted to these areas alone. Scientometrics can do well in adapting, testing and developing theories from various disciplines in the HSS. Glänzel and Schoepflin (1994) observe that not much has occurred in theoretical and methodological development. Among the disciplines of the HSS, there is no dearth of theories that can ably contribute to the growth and development of scientometrics, both theoretically and empirically. A major reason for the deficiency of scientometrics in not drawing much on theories from the HSS is that the volume of scientometric studies conducted in the HSS disciplines is small compared to that in science disciplines.

As demonstrated in the previous chapters, the status quo might change when more and more social scientists adopt scientometrics for the study of their disciplines in the HSS. With the use of rich qualitative data as described, the possibilities for a stronger theoretical basis are highly likely for scientometrics. Qualitative data can serve well for theoretical developments, i.e., grounded theories that originate from the data in scientometrics. Theories are developed from or through data generation in qualitative data analysis (Mason, 2002). Increased attention should be paid to this area for further advancements in scientometrics. It is also hoped that this book will lead to renewed interest in scientometric research, particularly in the HSS.

Notes

1 Gingras (2016) refers to a case in which over 80 per cent of the grants from the National Science Foundation in the US went to researchers cited more than 60 times on average in the previous five years. Similarly, the four academic departments that received the majority of the grants had researchers who were cited about 400 times in the same period.
2 Citation indexes have, however, been criticised for their use in evaluating scholars. The criticism emanates from the weaknesses of the mechanics and compiling of data, and the intrinsic character of the data. Garfield addressed these criticisms in his paper (Garfield, 1979).
3 In the beginning, it covered disciplines such as anthropology, archaeology, area studies, business and finance, communication, community health, criminology and penology, demography, economics, educational research, ethnic group studies, geography, history, information and library science, international relations, law, linguistics, management, marketing, philosophy, political science, psychiatry, psychology, sociology, statistics, and urban planning and development. In later years, more disciplines and subjects were added to it.
4 The disciplines include archaeology, architecture, art, classics, dance, film, TV and radio, folklore, history, language and linguistics, literature, music, oriental studies, philosophy, theatre, and theology and religious studies.
5 A research area which refers to both content and social aspects on which the focus falls on such studies, according to Small (2006:595), is a "set of documents or other bibliometric units that define a topic and an associated group of researchers who share an interest in the topic". Small's method comprised of adopting highly cited papers as the units of analysis, which is the top 1 per cent of the papers in each of the 22 broad disciplines. These units were later assembled into co-citation networks using clustering operations. These objects were then tracked to determine the pattern of continually highly cited papers at successive time periods.
6 Coauthorships of authors can be traced to different levels of collaboration – local, national, regional and/or international. Innumerable studies have been successfully conducted in

this area, applying scientometric methods. Even for a continent, as has been done for the whole of Africa (Sooryamoorthy, 2018, 2019, 2020), collaboration patterns can be ascertained.

7 It is important to mention some other databases that are useful in the study of the HSS. Archambault and Gagné (2004) compiled a list of useful databases that have the potential for scientometric research. Among them are America: History and Life (useful for history and culture); ABELL Online (literature, language and culture); CSA Worldwide Political Science Abstracts (political science); Francis (HSS); Historical Abstracts (history); Sociological Abstracts (sociology); Wilson Humanities Abstracts (humanities); and Wilson Social Science Abstracts (social sciences). Some of these are no longer available. DataCite has emerged as a source of scientometric data for the analysis and study of scholarly publications of open access data. Robinson-Garcia et al. (2017) have looked at the coverage, strengths and limitations of DataCite as a potential source of data. Kousha et al. (2018) used Microsoft Academic as a tool for assessing citation impact of in-press articles.

8 Web_of_Science_Core_Collection_long_guide_public_OCT-18, accessed 9 July 2019.

9 https://clarivate.com/products/web-of-science/databases/?utm_source=false&utm_medium=false&utm_campaign=false, accessed 19 April 2019.

10 This study was referring to 76,076 (2000–2015) peer-reviewed publications of researchers affiliated to a Flemish university.

11 For instance, the "Publication data collection instructions for researchers 2018" of the Finnish Ministry of Education and Culture (available from the web) provides a detailed list of the indicators and concepts that are useful in scientometric research.

References

Aaltojärvi, I., Arminen, I., Auranen, O. & Pasanen, H.-M. (2008). Scientific productivity, web visibility and citation patterns in sixteen Nordic sociology departments. *Acta Sociologica*, *51*, 5–22. https://doi.org/10.1177%2F0001699307086815

Adams, J. & Testa, J. (2011). Thomson Reuters Book Citation Index. In E. C. M. Noyons, P. Ngulube & J. Leta (Eds), *Proceedings of ISSI 2011: The 13th Conference of the International Society for Scientometrics and Informetrics* (pp. 13–18). Durban: ISSI.

Adriaanse, L. & Rensleigh, C. (2013). Web of Science, Scopus and Google Scholar. *The Electronic Library*, *31*, 727–744. https://doi.org/10.1108/EL-12-2011-0174

Aguillo, I. F. (2012). Is Google Scholar useful for bibliometrics? A webometric analysis. *Scientometrics*, *91*, 343–351. DOI: 10.1007/s11192-011-0582-8

Aksnes, D. W., Piro, F. N. & Rørstad, K. (2019). Gender gaps in international research collaboration: A bibliometric approach. *Scientometrics*, *120*, 747–774. https://doi.org/10.1007/s11192-019-03155-3

Andrés, A. (2009). *Measuring Academic Research: How to Undertake a Bibliometric Study*. Oxford: Chandos Publishing.

Archambault, É. & Gagné, É. V. (2004). *The Use of Bibliometrics in the Social Sciences and Humanities*. Montreal: Social Sciences and Humanities Research Council of Canada (SSHRCC). www.science-metrix.com/pdf/SM_2004_008_SSHRC_Bibliometrics_Social_Science.pdf

Archambault, É., Vignola-Gagné, É., Côté, G., Larivière, V. & Gingrasb, Y. (2006). Benchmarking scientific output in the social sciences and humanities: The limits of existing databases. *Scientometrics*, *68*, 329–342. https://doi.org/10.1007/s11192-006-0115-z

Ball, R. (2018). *An Introduction to Bibliometrics: New Development and Trends*. Cambridge, MA: Chandos Publishing.

Bar-Ilan, J. (2001). Data collection methods on the Web for informetric purposes – A review and analysis. *Scientometrics, 50*, 7–32. https://doi.org/10.1023/A:1005682102768

Bazeley, P. (2007). *Qualitative Data Analysis with NVivo*. London: Sage.

Björneborn, L. & Ingwersen, P. (2004). Toward a basic framework for webometrics. *Journal of the American Society for Information Science and Technology, 55*, 1216–1227. https://doi.org/10.1002/asi.20077

Blakeman, K. (2018). Bibliometrics in a digital age: Help or hindrance. *Science Progress, 101*, 293–310. https://doi.org/10.3184/003685018X15337564592469

Blank, G. (2004). Teaching qualitative data analysis to graduate students. *Social Science Computer Review*, 22, 187–196. https://doi.org/10.1177/0894439303262559

Bornmann, L. (2013). Is there currently a scientific revolution in scientometrics? *Journal of the Association for Information Science and Technology, 65*, 647–648. DOI: 10.1002/asi.23073

Bornmann, L. & Leydesdorff, L. (2014). Scientometrics in a changing research landscape. *EMBO Reports, 15*, 1228–1232. DOI: 10.15252/embr.201439608

Bornmann, L., Thor, A., Marx, W. & Schier, H. (2016). The application of bibliometrics to research evaluation in the humanities and social sciences: An exploratory study using normalized Google Scholar data for the publications of a research institute. *Journal of the American Society for Information Science and Technology, 67*, 2778–2789. https://doi.org/10.1002/asi.23627

Boschma, R. A. (2005). Proximity and innovation: A critical assessment. *Regional Studies, 39*, 61–74. https://doi.org/10.1080/0034340052000320887

De Bellis, N. (2009). *Bibliometrics and Citation Analysis: From the Science Citation Index to Cybermetrics*. Lanham, MD: The Scarecrow Press, Inc.

Frenken, K., Hardeman, S. & Hoekman, J. (2009). Spatial scientometrics: Towards a cumulative research program. *Journal of Informetrics, 3*, 222–232. https://doi.org/10.1016/j.joi.2009.03.005

Garfield, E. (1955). Citation indexes for science: A new dimension in documentation through association of ideas. *Science, 122*, 108–111. DOI: 10.1126/science.122.3159.108

Garfield, E. (1964). Citation indexing: A natural science literature retrieval system for the social sciences. *American Behavioral Scientist*, 7, 58–61. https://doi.org/10.1177/000276426400701017

Garfield, E. (1975). The Social Sciences Citation Index, more than a tool. *Current Contents, 12*, 6–9. www.garfield.library.upenn.edu/essays/v2p241y1974-76.pdf

Garfield, E. (1979). Is citation analysis a legitimate evaluation tool? *Scientometrics, 1*, 359–375. https://doi.org/10.1007/BF02019306

Garfield, E. (1983). How to use *Science Citation Index* (*SCI*). *Current Contents, 9*, 5–14. www.garfield.library.upenn.edu/essays/v6p053y1983.pdf

Garfield, E. (1984). How to use the Social Sciences Citation Index (SSCI). *Current Contents, 27*, 3–13. www.garfield.library.upenn.edu/essays/v7p202y1984.pdf

Garfield, E. (1985). How to use the *Arts & Humanities Citation Index* (*A&HCI*) and what's in it for you and your mate! *Current Contents, 6*, 3–12. www.garfield.library.upenn.edu/essays/v8p050y1985.pdf

Garfield, E. (1992). The uses and limitations of citation data as science indicators: An overview for students and nonspecialists. *Current Contents, 49*, 188. www.garfield.library.upenn.edu/essays/v15p188y1992-93.pdf

Garfield, E. (2007). The evolution of the *Science Citation Index*. *International Microbiology, 10*, 65–90. DOI: 10.2436/20.1501.01.10

Garfield, E. & Pudovkin, A. I. (2004). The HistCite system for mapping and bibliometric analysis of the output of searches using the ISI Web of Knowledge. Paper presented at the

Annual Meeting of ASIS&T, Newport, RI, 15 November. www.garfield.library.upenn. edu/papers/asist112004.pdf

Garfield, E., Pudovkin, A. I. & Istomin, V. S. (2002). Algorithmic citation-linked historiography: Mapping the literature of science. Paper presented at the Information, Connections and Community. 65th Annual Meeting of ASIST, Philadelphia, 18–21 November. https:// doi.org/10.1002/meet.1450390102

Garfield, E. & Welljarns-Dorof, A. (1992). Citation data: Their use as quantitative indicators for science and technology evaluation and policy-making. *Science and Public Policy, 19,* 321–327. https://doi.org/10.1093/spp/19.5.321

Gingras, Y. (2016). *Bibliometrics and Research Evaluation: Uses and Abuses.* Cambridge, MA: MIT Press.

Glänzel, W. (2008). Seven myths in bibliometrics about facts and fiction in quantitative science studies. *Collnet Journal of Scientometrics and Information Management, 2,* 9–17. https://doi. org/10.1080/09737766.2008.10700836

Glänzel, W. (2012). Bibliometric methods for detecting and analysing emerging research topics. *El profesional de la información, 21,* 194–201. DOI: 10.3145/epi.2012.mar.11

Glänzel, W. & Schoepflin, U. (1994). Little scientometrics, big scientometrics … and beyond? *Scientometrics, 30,* 375–384. https://doi.org/10.1007/BF02018107

Glänzel, W. & Schubert, A. (2004). Analysing scientific networks through co-authorship. In H. F. Moed, W. Glänzel & U. Schmoch (Eds), *Quantitative Science and Technology Research: The Use of Publication and Patent Statistics in Studies of S&T Systems* (pp. 257–276). New York: Kluwer Academic Publishers. DOI: 10.1007/1-4020-2755-9_12

Glänzel, W., Thijs, B. & Chi, P.-S. (2016). The challenges to expand bibliometric studies from periodical literature to monographic literature with a new data source: The Book Citation Index. *Scientometrics, 109,* 2165–2179. https://doi.org/10.1007/s11192-016-2046-7

Gómez-Morales, Y. J. (2015). Scientometrics. In G. Ritzer (Ed.), *The Blackwell Encyclopedia of Sociology.* London: John Wiley & Sons.

Gorraiz, J., Purnell, P. J. & Glänzel, W. (2013). Opportunities for and limitations of the Book Citation Index. *Journal of the American Society for Information Science and Technology, 64,* 1388–1398. https://doi.org/10.1002/asi.22875

Guns, R., Sīle, L., Eykens, J., Verleysen, F. T. & Engels, T. C. E. (2018). A comparison of cognitive and organizational classification of publications in the social sciences and humanities. *Scientometrics, 116,* 1093–1111. https://doi.org/10.1007/s11192-018-2775-x

Halevi, G., Moed, H. & Bar-Ilan, J. (2017). Suitability of Google Scholar as a source of scientific information and as a source of data for scientific evaluation – Review of the literature. *Journal of Informetrics, 11,* 823–834. DOI: 10.1016/j.joi.2017.06.005

Harding, J. (2013). *Qualitative Data Analysis: From Start to Finish.* London: SAGE.

Harzing, A.-W. K. (2014). A longitudinal study of Google Scholar coverage between 2012 and 2013. *Scientometrics, 98,* 565–575. DOI: 10.1007/s11192-013-0975-y

Harzing, A.-W. K. & Wal, R. v. d. (2008). Google Scholar as a new source for citation analysis. *Ethics in Science and Environmental Politics, 8,* 61–73. DOI: 10.3354/esep00076

Herrmannova, D. & Knoth, P. (2014). *Towards full-text based research metrics: Exploring semantometrics. Report of experiments.* Retrieved from https://pdfs.semanticscholar.org/ 7213/8af2afbf9713af6b56c441340ce398ea418b.pdf 17 October 2019.

Herrmannova, D. & Knoth, P. (2015). Semantometrics in coauthorship networks: Fulltext-based approach for analysing patterns of research collaboration. *D-Lib Magazine, 21,* www.dlib. org/dlib/november15/11contents.html. DOI: 10.1045/november2015-herrmannova

Holmberg, K. & Bowman, T. D. (2019). Google search results as an altmetrics data source? In G. Catalano, C. Daraio, M. Gregori, H. F. Moed & G. Ruocco (Eds), *Proceedings of the*

17th Conference of the International Society for Scientometrics and Informetrics (Vol. I, pp. 1462–1467). Rome: Edizioni Efesto.

Hood, W. W. & Wilson, C. S. (2003). Informetric studies using databases: Opportunities and challenges. *Scientometrics, 58*, 587–608. https://doi.org/10.1023/B:SCIE.000000 6882.47115.c6

Jacsó, P. (2005). As we may search – Comparison of major features of the Web of Science, Scopus, and Google Scholar citation-based and citation-enhanced databases. *Current Science, 89*, 1537–1547.

Jacsó, P. (2011). Google Scholar duped and deduped – The aura of "robometrics". *Online Information Review, 35*, 154–160. https://doi.org/10.1108/14684521111113632

Kent, R. (2015). *Analysing Quantitative Data: Variable-based and Case-based Approaches to Non-experimental Dataset*. London: SAGE.

Kousha, K., Thelwall, M. & Abdoli, M. (2018). Can Microsoft Academic assess the early citation impact of in-press articles? A multi-discipline exploratory analysis. *Journal of Informetrics, 12*, 287–298. DOI: 10.1016/j.joi.2018.01.009

Lamers, W. S., Eck, N. J. v. & Waltman, L. (2019). Variations in citation practices across the scientific landscape: Analysis based on a large full-text corpus. In G. Catalano, C. Daraio, M. Gregori, H. F. Moed & G. Ruocco (Eds), *Proceedings of the 17th Conference of the International Society for Scientometrics and Informetrics* (Vol. II, pp. 2121–2132). Rome: Edizioni Efesto.

Larivière, V., Archambault, É., Gingras, Y. & Vignola-Gagné, É. (2006). The place of serials in referencing practices: Comparing natural sciences and engineering with social sciences and humanities. *Journal of the American Society for Information Science and Technology, 57*, 997–1004. https://doi.org/10.1002/asi.20349

Larivière, V., Ni, C., Gingras, Y., Cronin, B. & Sugimoto, C. R. (2013). Bibliometrics: Global gender disparities in science. *Nature, 504*, 211–213. DOI: 10.1038/504211a

Leahey, E. (2006). Gender differences in productivity: Research specialization as a missing link. *Gender and Society, 20*, 754. https://doi.org/10.1177/0891243206293030

Leopold, E., May, M. & Paaß, G. (2004). Data mining and text mining for science & technology research. In H. F. Moed, W. Glänzel & U. Schmoch (Eds), *Quantitative Science and Technology Research: The Use of Publication and Patent Statistics in Studies of S&T Systems* (pp. 187–213). New York: Kluwer Academic Publishers. https://doi.org/10.1007/1-4020-2755-9_9

López-Cózar, E. D., Orduna-Malea, E., Martín, A. M.-. & Ayllón, J. M. (2018). Google Scholar: The big data bibliographic tool. In F. J. Cantú-Ortiz (Ed.), *Research Analytics: Boosting University Productivity and Competitiveness through Scientometrics* (pp. 59–80). Boca Raton, FL: CRC Press.

López-Cózar, E. D., Robinson-García, N. & Torres-Salinas, D. (2012). Manipulating Google Scholar citations and Google Scholar metrics: Simple, easy and tempting. *EC3 Working Papers*, Evaluación de la Ciencia y de la Comunicación Científica, No. 6.

Lyn, R. (1999). *Using NVivo in Qualitative Research*. London: Sage.

Ma, L. & Cleere, L. (2019). Comparing coverage of Scopus, WoS, and OBRSS list: A case for institutional and national databases of research output? In G. Catalano, C. Daraio, M. Gregori, H. F. Moed & G. Ruocco (Eds), *Proceedings of the 17th Conference of the International Society for Scientometrics and Informetrics* (Vol. I, pp. 214–222). Rome: Edizioni Efesto.

Martin, B., Tang, P., Morgan, M., Glänzel, W., Hornbostel, S., Lauer, G., … Zic-Fuchs, M. (2010). *Towards a Bibliometric Database for the Social Sciences and Humanities – A European Scoping Project*. A report produced for DFG, ESRC, AHRC, NWO, ANR and ESF. Brighton, UK: Science and Technology Policy Research Unit. https://globalhighered. files.wordpress.com/2010/07/esf_report_final_100309.pdf

Martin, B. R. (2011). What can bibliometrics tell us about changes in the mode of knowledge production? *Prometheus, 29,* 455–479. https://doi.org/10.1080/08109028.2011.643540

Martín-Martín, A., Orduna-Malea, E., Thelwall, M. & López-Cózar, E. D. (2019). Google Scholar, Web of Science, and Scopus: A systematic comparison of citations in 252 subject categories. *Journal of Informetrics, 12,* 1160–1177. https://doi.org/10.1016/j.joi.2018.09.002

Mason, J. (2002). *Qualitative Researching* (Second ed.). London: Sage.

Merigó, J., Pedrycz, W., Weber, R. & Sotta, C. d. l. (2018). Fifty years of Information Sciences: A bibliometric overview. *Information Sciences, 432,* 245–268. https://doi.org/10.1016/j.ins.2017.11.054

Miles, M. B. & Huberman, A. M. (1994). *Qualitative Data Analysis.* Thousand Oaks, CA: Sage.

Mingers, J. & Leydesdorff, L. (2015). A review of theory and practice in scientometrics. *European Journal of Operational Research, 246,* 1–19. https://doi.org/10.1016/j.ejor.2015.04.002

Moed, H. F. (1988). The use of on-line databases for bibliometric analysis. In L. Egghe & R. Rousseau (Eds), *Informetrics 87/88: Select Proceedings of the First International Conference on Bibliometrics and Theoretical Aspects of Information Retrieval, Diepenbeek, Belgium, 25–28 August 1987* (pp. 133–155). Amsterdam: Elsevier Science Publishers.

Neuhaus, C. & Daniel, H. D. (2008). Data sources for performing citation analysis: An overview. *Journal of Documentation, 64,* 193–210. DOI: 10.1108/00220410810858010

Prins, A. A. M., Costas, R., van Leeuwen, T. N. & Wouters, P. F. (2016). Using Google Scholar in research evaluation of humanities and social science programs: A comparison with Web of Science data. *Research Evaluation, 25,* 264–270. https://doi.org/10.1093/reseval/rvv049

Robinson-Garcia, N., Mongeon, P., Jeng, W. & Costas, R. (2017). DataCite as a novel bibliometric source: Coverage, strengths and limitations. *Journal of Informetrics, 11,* 841–854. https://doi.org/10.1016/j.joi.2017.07.003

Ruggunan, S. & Sooryamoorthy, R. (2016). Human resource management research in South Africa: A bibliometric study of authors and their collaboration patterns. *Journal of Contemporary Management, 13,* 1394–1427.

Ruggunan, S. & Sooryamoorthy, R. (2019). *Management Studies in South Africa: Exploring the Trajectory in the Apartheid Era and Beyond.* Basel, Switzerland: Springer.

Sangam, S. L. & Savanur, K. (2010). Eugene Garfield: A scientometric portrait. *Collnet Journal of Scientometrics and Information Management, 4,* 41–51. DOI: 10.1080/09737766.2010.10700883

Schmoch, U. (2019). Mean values of skew distributions in bibliometrics. In G. Catalano, C. Daraio, M. Gregori, H. F. Moed & G. Ruocco (Eds), *Proceedings of the 17th Conference of the International Society for Scientometrics and Informetrics* (Vol. I, pp. 160–166). Rome: Edizioni Efesto.

Schnell, J. D. (2018). Web of Science: The first citation index for data analytics and scientometrics. In F. J. Cantú-Ortiz (Ed.), *Research Analytics: Boosting University Productivity and Competitiveness through Scientometrics* (pp. 15–29). Boca Raton, FL: CRC Press.

Schotten, M., Aisati, M. e., Meester, W. J. N., Steiginga, S. & Ross, C. A. (2018). A brief history of Scopus: The world's largest abstract and citation database of scientific literature. In F. J. Cantú-Ortiz (Ed.), *Research Analytics: Boosting University Productivity and Competitiveness through Scientometrics* (pp. 31–58). Boca Raton, FL: CRC Press.

Science-Metrix. (2018). *Analytical Support for Bibliometrics Indicators: Development of Bibliometric Indicators to Measure Women's Contribution to Scientific Publications.* Montréal: Science-Metrix Inc.

Shapiro, F. R. (1992). Origins of bibliometrics, citation indexing, and citation analysis: The neglected legal literature. *Journal of the American Society for Information Science, 43,* 337–339. https://doi.org/10.1002/(SICI)1097-4571(199206)43:5<337::AID-ASI2>3.0.CO;2-T

Small, H. (2006). Tracking and predicting growth areas in science. *Scientometrics, 68*, 595–610. https://doi.org/10.1007/s11192-006-0132-y

Sooryamoorthy, R. (2015). Sociological research in South Africa: Post-apartheid trends. *International Sociology Reviews, 30*, 119–113. https://doi.org/10.1177/0268580915571801

Sooryamoorthy, R. (2016). *Sociology in South Africa: Colonial, Apartheid and Democratic Forms*. Hampshire and New York: Palgrave Macmillan.

Sooryamoorthy, R. (2018). The production of science in Africa: An analysis of publications in the science disciplines, 2000–2015. *Scientometrics, 115*, 317–349. https://doi.org/10.1007/s11192-018-2675-0

Sooryamoorthy, R. (2019). International collaboration in Africa: A scientometric analysis. In G. Catalano, C. Daraio, M. Gregori, H. F. Moed & G. Ruocco (Eds), *Proceedings of the 17th Conference of the International Society for Scientometrics and Informetrics* (Vol. I, pp. 151–159). Rome: Edizioni Efesto.

Sooryamoorthy, R. (2020). *Science, Policy and Development in Africa: Challenges and Prospects*. London: Cambridge University Press.

Srivastava, P. & Hopwood, N. (2009). A practical iterative framework for qualitative data analysis. *International Journal of Qualitative Methods, 8*, 76–84. https://doi.org/10.1177/160940690900800107

Sugimoto, C. R. & Larivière, V. (2018). *Measuring Research: What Everyone Needs to Know*. New York: Oxford University Press.

van Eck, N. J. & Waltman, L. (2010). Software survey: VOSviewer, a computer program for bibliometric mapping. *Scientometrics, 84*, 523–538. https://doi.org/10.1007/s11192-009-0146-3

van Eck, N. J. & Waltman, L. (2014). Visualizing bibliometric networks. In Y. Ding, R. Rousseau & D. Wolfram (Eds), *Measuring Scholarly Impact: Methods and Practice* (pp. 285–320). New York: Springer. https://doi.org/10.1007/978-3-319-10377-8_13

van Eck, N. J., Waltman, L., Dekker, R. & Van den Berg, J. (2010). A comparison of two techniques for bibliometric mapping: Multidimensional scaling and VOS. *Journal of the American Society for Information Science and Technology, 61*, 2405–2416. https://doi.org/10.1002/asi.21421

Visser, M., Eck, Nees Jan v. & Waltman, L. (2019). Large-scale comparison of bibliographic data sources: Web of Science, Scopus, Dimensions, and Crossref. In G. Catalano, C. Daraio, M. Gregori, H. F. Moed & G. Ruocco (Eds), *Proceedings of the 17th Conference of the International Society for Scientometrics and Informetrics* (Vol. II, pp. 2358–2369). Rome: Edizioni Efesto.

Wang, M. & Chai, L. (2018). Three new bibliometric indicators/approaches derived from keyword analysis. *Scientometrics, 116*, 721–750. https://doi.org/10.1007/s11192-018-2768-9

Williams, R. & Bornmann, L. (2016). Sampling issues in bibliometric analysis. *Journal of Informetrics, 10*, 1225–1232. https://doi.org/10.1016/j.joi.2015.11.004

Winter, J. C. F. d., Zadpoor, A. A. & Dodou, D. (2014). The expansion of Google Scholar versus Web of Science: A longitudinal study. *Scientometrics, 98*, 1547–1565. https://doi.org/10.1007/s11192-013-1089-2

Zavaraqi, R. & Fadaie, G.-R. (2012). Scientometrics or science of science: Quantitative, qualitative or mixed one. *Collnet Journal of Scientometrics and Information Management, 6*, 273–278. https://doi.org/10.1080/09737766.2012.10700939

INDEX

Note: *Scientometrics for the Humanities and Social Sciences*

For Product Safety Concerns and Information please contact our EU
representative GPSR@taylorandfrancis.com
Taylor & Francis Verlag GmbH, Kaufingerstraße 24, 80331 München, Germany